World Scientific Series in Current Energy Issues　Volume 7

Wind, Water and Fire
The Other Renewable Energy Resources

World Scientific Series in Current Energy Issues

Series Editor: Gerard M Crawley *(University of South Carolina & Marcus Enterprise LLC, USA)*

World Scientific Series in Current Energy Issues | Volume 7

Wind, Water and Fire

The Other Renewable Energy Resources

Editor

Gerard M Crawley

Marcus Enterprises LLC, USA

World Scientific

NEW JERSEY · LONDON · SINGAPORE · BEIJING · SHANGHAI · HONG KONG · TAIPEI · CHENNAI · TOKYO

Published by

World Scientific Publishing Co. Pte. Ltd.

5 Toh Tuck Link, Singapore 596224

USA office: 27 Warren Street, Suite 401-402, Hackensack, NJ 07601

UK office: 57 Shelton Street, Covent Garden, London WC2H 9HE

Library of Congress Cataloging-in-Publication Data

Names: Crawley, Gerard M., editor.

Title: Wind, water and fire : the other renewable energy resources / editor, Gerard M. Crawley, Marcus Enterprises LLC, USA.

Description: New Jersey : World Scientific, [2021] | Series: World scientific series in current energy issues, 2425-0163 ; Vol. 7 | Includes bibliographical references and index.

Identifiers: LCCN 2020038664 (print) | LCCN 2020038665 (ebook) | ISBN 9789811225918 (hardcover) | ISBN 9789811225925 (ebook) | ISBN 9789811225932 (ebook other)

Subjects: LCSH: Renewable energy sources.

Classification: LCC TJ808 .W57 2021 (print) | LCC TJ808 (ebook) | DDC 333.79/4--dc23

LC record available at https://lccn.loc.gov/2020038664

LC ebook record available at https://lccn.loc.gov/2020038665

British Library Cataloguing-in-Publication Data

A catalogue record for this book is available from the British Library.

For any available supplementary material, please visit
https://www.worldscientific.com/worldscibooks/10.1142/11985#t=suppl

Desk Editors: Balamurugan Rajendran/Amanda Yun

Typeset by Stallion Press
Email: enquiries@stallionpress.com

Foreword to the World Scientific Series on Current Energy Issues

Sometime between four hundred thousand and a million years ago, an early humanoid species developed the mastery of fire and changed the course of our planet. But, as recently as a few hundred years ago, the energy sources available to the human race remained surprisingly limited. In fact, until the early nineteenth century, the main energy sources for humanity were biomass (from crops and trees), their domesticated animals and their own efforts.

Even after many millennia, the average per capita energy use in 1830 only reached about 20 Gigajoules[a] (GJ) per year. By 2010, however, this number had increased dramatically to about 80 GJ per year.[1] One reason for this notable shift in energy use is that the number of possible energy sources increased substantially during this period, starting with coal in about the 1850s and then successively adding oil and natural gas. By the middle of the twentieth century, hydropower and nuclear fission were added to the mix. As we move into the twenty-first century, there has been a steady increase in other forms of energy such as wind and solar, although presently they represent a relatively small fraction of world energy use.

Despite the rise of a variety of energy sources, per capita energy use is not uniform around the world. There are enormous differences from country to country, pointing to a large disparity in wealth and opportunity (See Table 1). For example, in the United States the per capita energy use per year in 2017 was 301.2 million Btu[2] (MMBtu) and in Germany, 169.5 MMBtu. In China, however, per capita energy use was only

---

[a]Note 1 GJ = 0.947 MMBtu.

Table 1.   Primary Energy Use per capita in million Btu (MMBtu).[2]

Country	2007 (MMBtu)	2017 (MMBtu)	Percentage Change
Canada	416.1	411.7	−1.1%
United States	336.9	301.2	−10.6%
Brazil	52.7	60.4	+14.6%
France	175.7	154.1	−12.3%
Germany	167.8	169.5	+1.0%
Russia	204.0	275.6	+35.1%
Nigeria	6.1	8.1	+32.8%
Egypt	36.4	41.6	+14.3%
China	57.1	138.7	+142.9%
India	17.0	22.7	+33.5%
World	**72.2**	**77.3**	**+7.1%**

57.1 MMBtu in 2007, but grew dramatically to 138.7 MMBtu in 2017. India also saw a significant increase in per capita energy use from 2007 to 2017 of 33.5%. The general trends over the last decade suggest that countries with developed economies generally show modest increases or even small decreases in energy use, but that many developing economies, particularly China and India, are experiencing rapidly increasing energy consumption per capita.

These changes, both in the kind of energy resource used and the growth of energy use in countries with developing economies, will have enormous effects in the near future, both economically and politically, as greater numbers of people compete for limited energy resources at a viable price. A growing demand for energy will have an impact on the distribution of other limited resources such as food and fresh water as well. All this leads to the conclusion that energy will be a pressing issue for the future of humanity.

All energy sources have disadvantages as well as advantages, risks as well as opportunities, both in the production of the resource and in its distribution and ultimate use. Coal, the oldest of the "new" energy sources, is still used extensively to produce electricity, despite its potential environmental and safety concerns in both underground and open cut mining. Burning coal releases sulfur and nitrogen oxides which in turn can lead to acid rain and a cascade of detrimental consequences. Coal production requires careful regulation and oversight to allow it to be used safely and without damaging the environment. Even a resource like wind energy, which uses large wind turbines, has its critics because of the potential for bird kill and noise pollution. Some critics also find large wind turbines an unsightly addition to the landscape, particularly when the wind farms are erected in pristine environments. Energy from nuclear fission, originally believed to be "too cheap to meter"[3] has not had the growth predicted because of the problem with long-term storage of the waste from nuclear reactors and because of the public perception regarding the danger of catastrophic accidents such as happened at Chernobyl in 1986 and at Fukushima in 2011.

Even more recently, the amount of carbon dioxide in the atmosphere has steadily increased and is now greater than 400 parts per million (ppm).[4] The amount of another gas, methane (CH_4), in the atmosphere is also increasing and it is an even more potent greenhouse gas than CO_2. Methane is often released as part of the extraction of oil. The increase of these greenhouse gases has raised concerns in the scientific community about the continued use of fossil fuels and has led the majority of climate scientists to conclude[5] that this will result in a significant increase in global temperatures. We will see a rise in ocean temperature, acidity and sea level, all of which will have a profound impact on human life ands ecosystems around the world. Relying primarily on fossil fuels into the future may therefore prove precarious, since burning coal, oil and natural gas will necessarily increase CO_2 levels. Certainly, for the long term, adopting a variety of alternative energy sources which do not produce CO_2 nor release methane, seems to be our best strategy.

In addition, we should consider ways to use energy more efficiently, including better insulation of our buildings, more energy efficient manufacturing and much more energy-efficient modes of transportation. As noted in Table 1, there remains as much as a factor of two in energy use per capita even among developed economies. As energy becomes more expensive in the future, this will undoubtedly provide additional incentives for more efficient energy use.

The volumes in the World Scientific Series on Current Energy Issues explore different energy resources and issues related to the use of energy. The volumes are intended to be comprehensive, accurate, current and include an international perspective. The authors of the various chapters are experts in their respective fields and provide reliable information that can be useful to scientists and engineers, but also to policy makers and the general public interested in learning about the essential concepts related to energy. The volumes will deal with the technical aspects of energy questions, but will also include relevant discussion about economic and policy matters.

The goal of the series is not polemical, but rather is intended to provide information that will allow the reader to reach conclusions based on sound, scientific data.

The role of energy in our future is critical and will become increasingly urgent as the world's population increases and the global demand for energy turns ever upwards. Questions such as which energy sources to develop, how to store energy and how to manage the environmental impact of energy use will take centerstage in our future. The distribution and cost of energy will have powerful political and economic consequences and must also be addressed. How the world deals with these questions will make a crucial difference to the future of the Earth and its inhabitants. Careful consideration of our energy use today will have lasting effects for tomorrow. We intend that the World Scientific Series on Current Energy Issues will make a valuable contribution to this discussion.

References

1. Our Finite World: World energy consumption since 1820 in charts. March 2012. Accessed in February 2015 at http://ourfiniteworld.com/2012/03/12/world-energy-consumption-since-1820-in-charts/
2. U.S. Energy Information Administration, Independent Statistics & Analysis. Accessed in February 2020 at http://www.eia.gov/cfapps/ipdbproject/iedindex3.cfm?tid=44&pid=45&aid=2&cid=regions&syid=2005&eyid=2011&unit=MBTUPP
3. www.nrc.gov/docs/ML1613/ML1613A120.pdf. Remarks prepared by Lewis L. Strauss, Chairman, US Atomic Energy Commission, 16th Sep, 1954. Accessed in February 2020. There is some debate as to whether Strauss actually meant energy from nuclear fission or not.
4. NOAA Earth System Research Laboratory, Trends in Atmospheric Carbon Dioxide. Accessed in March 2015 at http://www.esrl.noaa.gov/gmd/ccgg/trends/
5. IPCC, Intergovernmental Panel on Climate Change, Fifth Assessment report 2014. Accessed in March 2015 at http://www.ipcc.ch/

Contents

Chapter 3. Costing and Future Development of Wind
Turbine Technology 63

Li Zhang and Qiou Yang Zhou

Chapter 4. Geothermal Energy 85

Magnus Gehringer

Chapter 7. Wave Energy   197

Arne Vögler

Chapter 8. Tidal Power   233

Sue Molloy

Chapter 1

Wind Energy Development: History and Current Status

Li Zhang

The University of Leeds, Leeds, UK
l.zhang@leeds.ac.uk

Abstract

Wind energy is free, renewable and sustainable, and it is becoming one of the dominant and fastest-growing renewable energy technologies in the world. Electricity can be produced from the wind without causing any environmental pollution. This chapter starts by giving a brief review of wind energy utilization by mankind from the Middle Ages till the late 20th century, leading on to the wind turbines developed to generate electricity and supply it to the power grid. The current status of wind converter development worldwide is then presented, covering the global cumulative installed capacity of wind power up to 2017, in both the onshore and offshore wind plants. The factors that promote the development of wind projects are discussed in the context of countries at the forefront of wind energy development.

1 Wind Energy: Nature and Origin of Development

Wind is a motion of a mass of air, blowing across the Earth from high-pressure areas to low-pressure areas, and bringing with it inexhaustible natural motive power. Mankind has used wind energy for thousands of years, with early applications including pumping water and grinding grains. The construction techniques of the time discovered in the Middle East were mainly based on using vertical axes to extract drag force from the wind. Early use of wind energy in Europe dates back to 1191 AD, when countries with rich wind resources, such as Holland and England, built horizontal axis windmills with tilted wings or sails. By the 18th century AD, multi-sail Dutch windmills were extensively used in Europe, with numbers estimated at about 8000 windmills in operation in Holland and 10,000 in England and Germany around 1750 AD. European settlers also built windmills in North America, mainly along the east coast area. The use of wind power for sailing ships and the mastering of sea navigation brought enormous fortunes to the European colonial powers until the age of steam. Then with the intermittent nature and uncertain availability of wind combined with the slowness of wind powered vessels, these gradually gave way to fossil-powered commercial shipping. Nevertheless, yachting and small boat sailing remain important recreational sports throughout the world.

The development of wind turbines with aerodynamically formed rotor blades for electricity generation only started around 80 years ago, including projects by German engineers (Kleinhenz and Honnef), and by the American Palmer Putnam (for the Smith company) in the 1940s which produced a turbine of 1250 kW. Wind powered machines were mass produced in the early 1950s, mainly by the German construction company Allgaier, to supply electricity to farms lying far from the public grid. However, these were small-scale generators having power ratings of 6 kW to 10 kW. Large megawatt class wind turbines re-appeared only in the 1970s due to rising fuel prices. Particularly in the USA, Sweden and Germany, megawatt machines were produced such as the American MOD-2 with 2.5 MW nominal power and 91 m rotor diameter, the 61 m Swedish-American 4 MW WTS-4, Swedish 3 MW WTS-75 AEOLUS model, and the German 3 MW GROWIAN having a 100 m rotor diameter and 100 m hub height.

Independently of large turbine development, the use of wind-generated power to supply energy to the grid on a large scale started in the 1980s in the USA state of California. Other countries including Denmark, Holland and Germany were also developing grid connected wind converters. Turbines of initially 50 kW, steadily scaled up to megawatt ranges, were used. This development has led to mass production of wind turbines and has resulted in a considerable improvement in performance.

2   Current Status of Wind Converter Development WorldWide

Wind power-based renewable energy is the fastest growing energy technology in the world. The rate of development and deployment accelerated

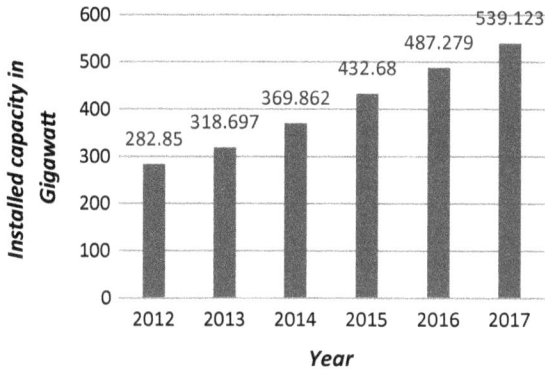

Fig. 1.   Global cumulative installed capacity of wind power from 2012 to 2017.[1]

rapidly over the last decade. From 2000 to 2016, the cumulative installed wind capacity increased at a compound annual rate of 15%. Figure 1 shows[1] the change in global wind power capacity during the period between 2012 and 2017.

By the end of 2012, around 180,000 wind turbines with the total capacity of 282 GW had been installed globally. Additionally, about 900,000 small wind turbines with power ratings below 100 kW were in operation with a total estimated capacity of 850 MW. Then in 2015, a record high amount of new wind capacity, roughly 63,000 MW, was added, yielding a cumulative total of around 433,000 MW.

By the end of 2016, the total installed wind capacity had reached 487 GW, with 454 GW onshore. China accounted for 38% of new annual capacity additions in 2016, followed by the United States (17%), Germany (10%), India (7%), Brazil (4%) and France (3%).[2] Net additions of wind power were 21% lower in 2016 than in 2015, the record year in which 65 GW was added to global capacity. This was mainly due to policy changes in China, which drove a rush of installation before the expiration of a policy support scheme at the end of 2015. China added 42% less capacity in 2016 than in 2015, accounting for almost all the global difference between 2016 and 2015. The range of expected yearly additions in the next 3–5 years is 40–50 GW. China, the United States, Germany, India and France are expected to account for the majority of new additions.[3]

All the above rapid growth is fanned on the one hand by the urge to reduce greenhouse gas emissions worldwide and the desire to solve the world energy shortage problem. It is also supported by government incentive

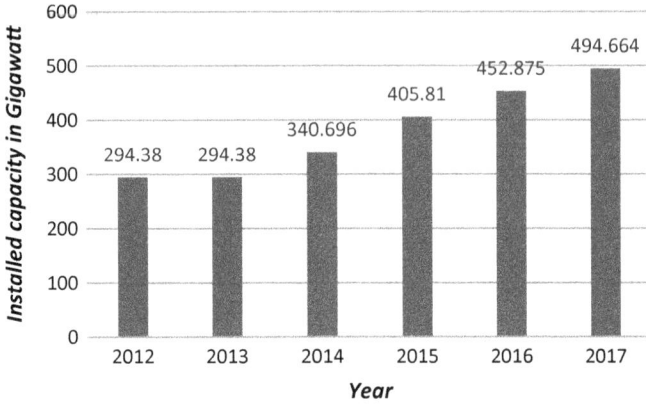

Fig. 2. The cumulative installed capacity of onshore wind farms between 2012 and 2017.[4]

policies, improvements in the cost and performance of wind power technologies, yielding low power sales prices for utilities, corporates and other purchasers.

2.1 *Onshore wind plants*

Figure 2 shows the development trend of onshore wind farms from 2012 to 2017. By the end of 2016, the capacity of onshore wind farm worldwide had reached 494 GW. An upward trend could be found in the global cumulative capacity of onshore wind energy from this statistic, especially in the last four years, the wind turbine capacity had increased at a rate of 50 GW per year. Figure 3 shows the cumulative installed capacity of wind power for leading countries by the end of 2017.

China has once more underpinned its role as the global wind power leader, occupying a market share of approximately 33%. In addition, during this year, the capacity of onshore wind farms in Germany was over 50 GW. In 2016, in Europe alone, there was 168.8 GW installed capacity of wind farms consisting of 153 GW onshore wind farms and 15.8 GW offshore. Three countries are the main contributors, namely, Germany, the United Kingdom and Spain, and have around four-fifths of Europe's total installed capacity. Moreover, the United Kingdom government is aiming to generate 20% of the country's electricity from wind turbines by 2020, up from 3% in 2008.

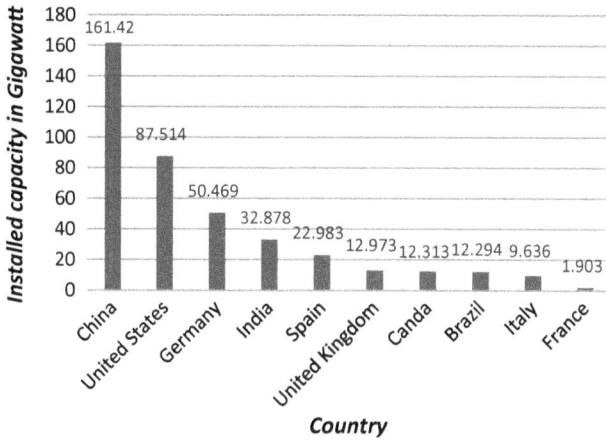

Fig. 3. The cumulative installed capacity of onshore wind farms for leading countries in 2017.[5]

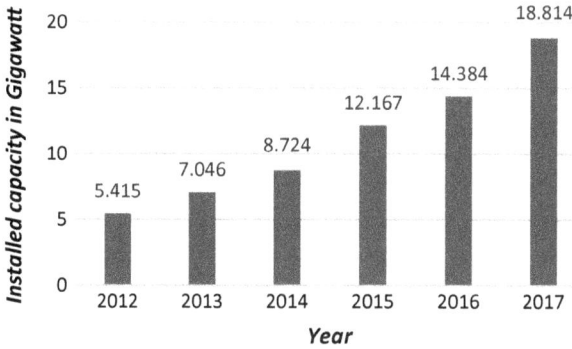

Fig. 4. Global cumulative installed capacity of offshore wind farm from 2012 to 2017.[6]

2.2 Offshore wind plants

There are many factors that promote the development of offshore wind projects, including the improvement and maturation of the technology, better management, growing investor confidence and the introduction and deployment of a new generation of turbines, with enormous swept area and tremendous output. The cumulative offshore wind farm capacity is shown in Fig. 4.

The record shows that, at the end of 2017, nearly 84% (15,780 MW) of all offshore installations were located in the waters off the coast of

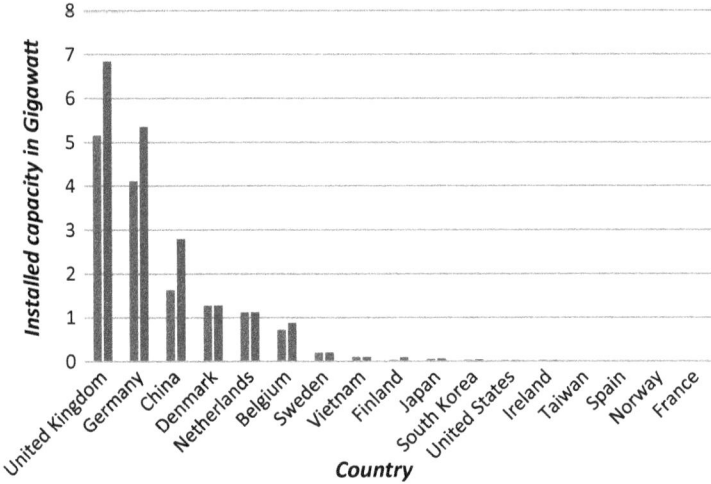

Fig. 5. The cumulative installed capacity of offshore wind farm among leading countries in 2016 and 2017.[6]

11 European countries. The remaining 16% is located largely in China, followed by Vietnam, Japan, South Korea, the United States and Taiwan.

As shown in Fig. 5,[6] the United Kingdom is the world's largest offshore wind market and accounts for over 36% of installed capacity, followed by Germany in the second spot with 28.5%. China comes third in the global offshore rankings with just under 15%. Denmark now accounts for 6.8%, the Netherlands 5.9%, Belgium 4.7% and Sweden 1.1%. Other countries including Vietnam, Finland, Japan, South Korea, the US, Ireland, Taiwan, Spain, Norway and France make up the balance of the market.

References

1. https://www.statista.com/statistics/268363/installed-wind-power-capacity-worldwide/.
2. http://www.irena.org/media/Files/IRENA/Agency/Publication/2018/Jan/IRENA_2017_Power_Costs_2018.pdf.
3. http://gwec.net/global-figures/graphs/.
4. https://www.statista.com/statistics/476306/global-capacity-of-onshore-wind-energy/.
5. https://www.statista.com/statistics/476318/global-capacity-of-onshore-wind-energy-in-select-countries/.
6. http://gwec.net/global-figures/graphs/.

Chapter 2

Electricity Generation by Wind Energy

Li Zhang

The University of Leeds, Leeds, UK
l.zhang@leeds.ac.uk

Abstract

The aim of this chapter is to give the non-specialist a comprehensive appreciation of all the main technical aspects of electricity generation from wind power. The first section treats basic aerodynamic theory of wind energy conversion, with key equations showing how the output depends on wind speed, and the limiting conversion efficiencies that can be achieved. The second section discusses the technology of common wind power generation systems, introducing the principles of synchronous and induction machines and their application in approaches to generation based on fixed or variable speed and frequency. Any of these should form a summary and a basis for further study, and should help graduating engineers or researchers, newly entering the field, to appreciate their work in its general context.

1 Part 1: Power and Energy Basis of Wind Converters

The term wind energy or wind power describes the process by which the wind is used to generate electrical power or electricity. Wind turbines convert the kinetic energy in the wind into mechanical power.[1] Nearly all modern wind turbines for electricity generation are horizontal axis machines with fast running rotors, making use of lifting force for higher efficiency. Thus all the analysis in this chapter will be based on this type of machine.

Components of a Horizontal Axis Wind Turbine

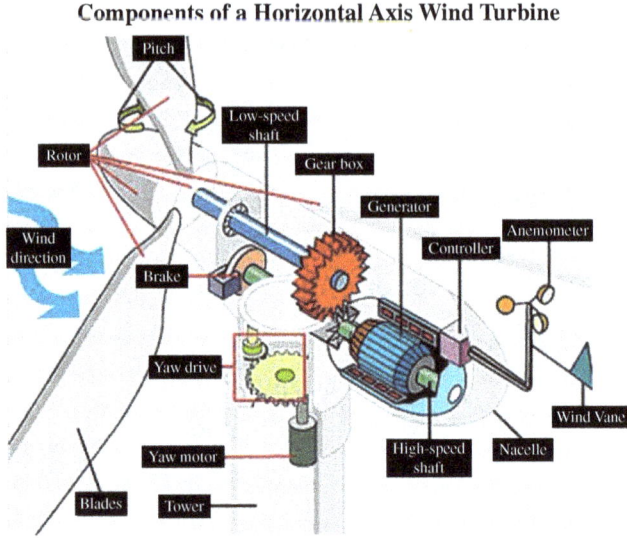

Fig. 1.   Components in a horizontal axis wind turbine.[2]

1.1   *Horizontal axis wind turbine* (*HAWT*)

The principal components in a modern horizontal axis wind turbine are shown in Fig. 1 and described as follows:

- The rotor blades, which extract the kinetic energy present in the wind and transform it into mechanical power;
- The nacelle, with a power control system that limits and conditions the extracted power;
- A gearbox that transfers the load and increases the rotational speed to drive the generator;
- An electrical system which converts the mechanical energy into electrical energy and
- A tower that supports the nacelle.

1.2   *Power extractable from the airstream*

From the viewpoint of energy conversion, the most important properties of the wind at a particular location are the velocity of the airstream and the air density. The power density ρ varies with altitude and with

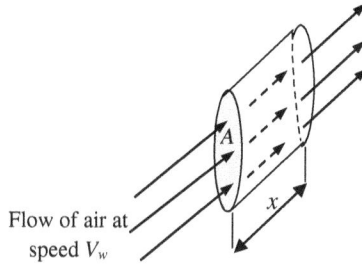

Fig. 2. Flow of air through space with area A and thickness x.

atmospheric conditions, such as temperature, pressure and humidity. At sea level and at standard temperature and pressure (STP), the value is $\rho = 1.202\,\text{kg/m}^3$ at 1,000 millibars (29.53 inches of mercury) or 101.3 kilopascal (kPa) pressure, and temperature 293 K. The temperature, pressure and density of air decrease with altitude. For wind-turbine applications, the range of interest is mostly within a couple of hundred feet of ground level.

If a smooth and laminar flow of a wind mass at an average velocity V_w, passes perpendicularly through an area A of any shape with thickness x, as shown in Fig. 2, the mass of air is $m = \rho A x$ and its kinetic energy is

$$KE = \frac{1}{2}\rho A x V_w^2.$$

A fresh block of air could be processed after the time x/Vw taken for the first block to cover the distance x, so if all the kinetic energy could be extracted, the power in the wind appears to be

$$P_W = \frac{1}{2}\rho A V_w^3. \qquad (1)$$

Equation (1) gives the most important feature, namely, the wind power is proportional to the cube of the average wind speed, which is the basis of all wind power and energy calculations. However, only a fraction of the total theoretical power available in the wind is extractable, due to various energy losses.

Consider the energy absorption from a flow of smooth and steady air of volume V_a with an upstream average velocity V_1 impinging upon a turbine rotor, and flowing far downstream at a smaller average velocity V_2, as shown

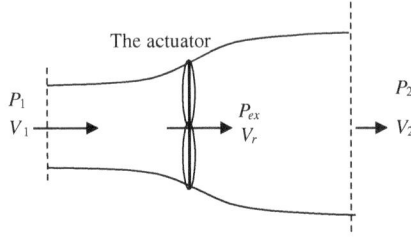

Fig. 3. Power extractions from the wind converter.

in Fig. 3, it is

$$KE = \frac{V_a}{2}\rho(V_1^2 - V_2^2).\tag{2}$$

The power extracted by the wind turbine may be

$$P_w = \frac{d}{dt}(KE) = \frac{d}{dt}\left(\frac{V_a}{2}\rho(V_1^2 - V_2^2)\right).\tag{3}$$

The rate of change of an air volume V_a relates to the product of effective area A and the wind speed V_r at the rotor surface,

$$\frac{dV_a}{dt} = AV_r.\tag{4}$$

Substituting (4) into (3) gives

$$P_w = \frac{1}{2}\rho AV_r(V_1^2 - V_2^2).\tag{5}$$

In the process of extracting energy from the wind, the wind velocity V_r that actuates the rotor is less than the upstream "free wind" velocity V_1. With an ideal and lossless system, all of the energy reduction in the airstream is transferred to the rotor of the wind machine. The downstream wind velocity V_2 is smaller than V_r.

The air mass passing through the rotor undergoes not only an energy reduction but a reduction of linear momentum:

$$\text{Reduction of momentum} = \rho Ax(V_1 - V_2).$$

The time rate of the change of momentum reduction is a force of value:

$$\frac{d}{dt}(\rho Ax(V_1 - V_2)) = \rho AV_r(V_1 - V_2).\tag{6}$$

By equating the rate of the change of kinetic energy transfer in Eq. (3) with the power associated with the rate of change of momentum, $V_r = \frac{V_1+V_2}{2}$. Thus the wind velocity at the rotor is the average of the upstream and downstream steady wind velocities. Substituting V_r into (6) gives the expression for the power extractable from the wind by the rotor as

$$P_{ex} = \frac{1}{2}\rho A V_r (V_1^2 - V_2^2) = \frac{1}{4}\rho A (V_1 + V_2)(V_1^2 - V_2^2)$$

$$= \frac{1}{4}\rho A V_1^3 \left[1 + \frac{V_2}{V_1} - \left(\frac{V_2}{V_1}\right)^2 - \left(\frac{V_2}{V_1}\right)^3\right]$$

$$= \frac{1}{4}\rho A V_1^3 \left[1 + \frac{V_2}{V_1}\right]\left[1 - \left(\frac{V_2}{V_1}\right)^2\right]. \tag{7}$$

The wind ratio V_2/V_1 that results in the maximum power transfer is calculated by differentiating (7) w.r.t V_2/V_1 and equating the resultant expression to zero. The valid result $V_2/V_1 = 1/3$ is taken and substituted into (7) to give an expression for the maximum possible power that can be extracted under ideal conditions,

$$P_{ex}(\text{max}) = \frac{8}{27}\rho A V_1^3 = 0.593 \cdot \frac{1}{2}\rho A V_1^3. \tag{8}$$

This very important result is sometimes referred to as Betz Law,[3] being named after the German scientist Albert Betz of Gottingen. This states that even with ideal wind energy conversion, the maximum power transferable is only 0.593 of the total power in the wind. In reality, only a fraction of this theoretical maximum power is obtained even by the best designed turbines.

1.2.1 *Practical extractable power and power coefficient C_p*

The power extractable by a practical wind turbine is much less than the maximum theoretical value defined in Eq. (7). A practical wind machine experiences air drag and friction on the rotor blades causing power to be lost in heat generation. In addition, the rotation of the blades causes swirling of the air, which reduces the torque imparted to the blades. The net effect of various losses is incorporated into a parameter called the power coefficient C_p.

With an upstream velocity V_1, the extractable power is:

$$P_{ex} = C_p \frac{1}{2}\rho A V_1^3, \tag{9}$$

so accordingly

$$C_p = \frac{1}{2}\left[1 + \frac{V_2}{V_1}\right]\left[1 - \left(\frac{V_2}{V_1}\right)^2\right]. \tag{10}$$

For practical wind turbines, the value varies in the range $0 \leq C_\nu \leq 0.4$. With the typical value $C_p = 0.4$, for example, it is seen that power available in the wind is $0.4/0.593$ or about 67% of the ideal theoretical value.

The value of C_p depends on the wind velocity, turbine rotational velocity and turbine blade parameters such as pitch angle.

1.2.2   Tip-speed ratio (TSR)

In order to express the power coefficient C_p in terms of both the upstream wind velocity V_1 and the blade rotational angular velocity ω_t, a parameter called the tip-speed ratio (TSR) is defined as:

$$\text{TSR} = \lambda = \frac{r\omega_t}{V_W} = \frac{v_t}{V_w}, \tag{11}$$

where v_t is the instantaneous velocity in meter/second of the turbine blade tip, tangential to its rotational motion, and ω_t is the turbine angular velocity in radians per second. With rotational speed n_t (r.p.m.), $\omega_t = \frac{2\pi n_t}{60}$. For a fixed blade pitch angle, for every value of ω_t, the characteristic of C_p against V_w is different. Likewise, for every V_w, C_p against ω_t is different. The characteristic of C_P versus TSR is a "universal" curve that subsumes values of both ω_t and V_w. Figure 4 shows a more detailed performance characteristic for Darrieus propeller machines; the peak value of C_p approaches 0.45, which is typical of small wind converters.

Good rotor design requires that the maximum value of the power coefficient C_{Pm} occurs near the design-rated value of the rotational speed.

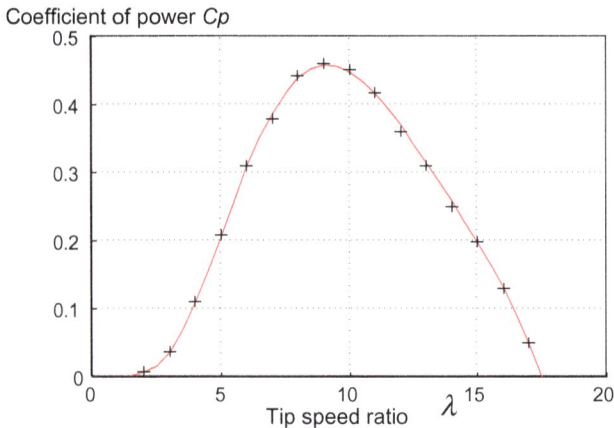

Fig. 4.   Power coefficient C_p vs TSR for Darrieus Propeller turbines.[3]

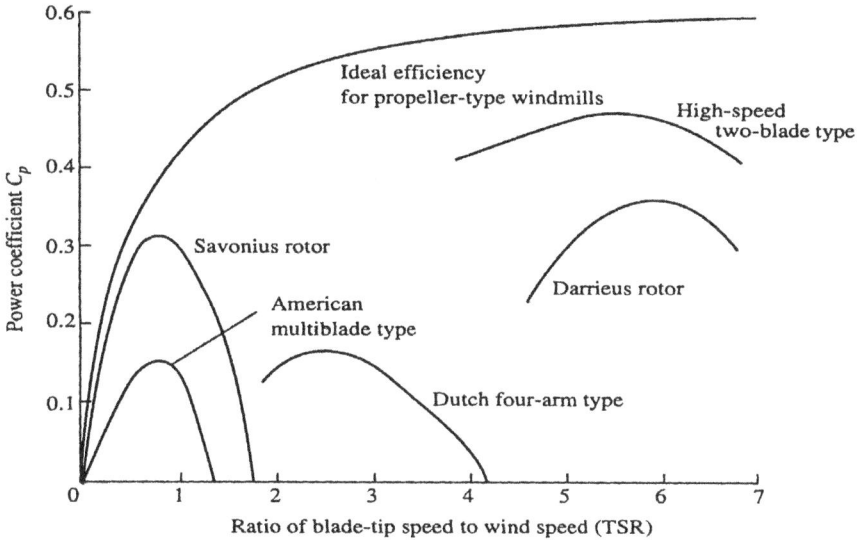

Fig. 5. C_p-TSR for different wind turbines.[4]

Typical characteristics for various different types of wind turbines are shown in Fig. 5. The maximum ideal efficiency characteristic for propeller machines is asymptotic to the Betz Law value of 0.593. It can be seen that the most efficient forms of wind converter are the propeller types for which $0.4 \leq C_{Pm} \leq 0.5$. In addition, the maximum value of the power coefficient is designed to occur in a certain range of TSR, namely $4 < TSR < 7$.

1.2.3   *Forces on turbine blades and control*

The cross-section of a wind turbine rotor blade is designed like the cross-section of an airplane wing (or aerofoil) to give the maximum transfer of energy. The aerofoil has a different camber on its two surfaces. Air passing over the top of the aerofoil has a higher velocity than on the lower surface, and hence lower pressure, by Bernoulli's principle. This results in a lifting force F_{lift} perpendicular to the airflow V_r, and drag force F_{drag} parallel to the flow. There is also a torque produced by the two forces, called the pitch moment.

   Figure 6 shows airflow velocities and forces on a rotor blade segment. An imaginary line from the rounded leading edge to the tapered trailing edge is called the wing chord. V_w indicates the wind velocity, which is parallel to the rotor axis and perpendicular to the plane of rotation. V_r is

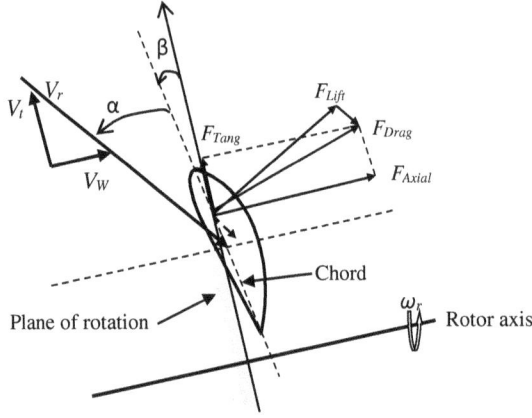

Fig. 6.   Airflow velocities and forces on a turbine blade segment.[6]

the resultant wind velocity parallel to the plane of rotation. The angle between the wing chord line and V_r is called the angle of attack α. Two forces acting directly on the turbine blade, the drag force and lift force, are shown.

Depending on the blade radius r, there is a different flow behavior at the profile for different blade angles.

1.2.3.1   Drag force

When a turbine blade is set up perpendicular to the wind, a drag force F_D acts on the blade in the direction of the resultant wind that can be expressed as[5]

$$F_D = c_D(\alpha)\frac{1}{2}\rho A_v v_r^2,\tag{12}$$

where v_r expresses the resultant wind velocity, the effective sweep area $A_v = t \times r$ is the projected body area of a rotor blade, approximately equal to rotor radius r and chord length t. The drag coefficient is $c_D(\alpha)$, which is dependent on the shape of the blade. The angle of attack, α, is that between v_r and the chord of the aerofoil.

1.2.3.2   Lift force

This acts on a blade due to the wind circulating around it. It is perpendicular to the drag force and this develops higher flow speed along the upper surface than along the lower; an overpressure emerges at the lower surface and an underpressure at the upper. A lift force is given by[5]

$$F_L = c_L(\alpha) \frac{1}{2} \rho A_v v_r^2, \tag{13}$$

where $c_L(\alpha)$ is the lift coefficient.

The total force is the vector sum of drag and lift forces $\bar{F}_R = \bar{F}_D + \bar{F}_L$. The resultant force can be divided into an axial component $\bar{F}_{Axial}$ and a tangential component $\bar{F}_{Tang}$ as shown in Fig. 6. The latter causes the rotor to turn. Both forces are dependent on the following:

- The pitch angle β of the blade element; this is the chord with respect to the plane of rotation,
- The angle of attack α; this is the angle between the resultant wind velocity V_r and the chord of the aerofoil,
- The lift and drag coefficients c_L and c_D for the blade profile.

Thus, the tangential and axial forces can be expressed, respectively, as[4]

$$F_{Tang} = \frac{1}{2} \rho A_v v_r^2 (c_L \cos \beta - c_D \sin \beta) \tag{14}$$

$$F_{Axail} = \frac{1}{2} \rho A_v v_r^2 (c_L \sin \beta - c_D \cos \beta). \tag{15}$$

For the respective wind and blade tip velocity, the tangential force is essentially dependent on the blade pitch angle, β, and changes of these two power coefficients, hence the total C_p.

Thus, the power and torque extractable from the wind can be

$$P_{ex} = C_P(\lambda, \beta) \frac{1}{2} \rho A_v v_r^3 \quad \text{and} \tag{16}$$

$$T_{ex} = \frac{P_{ex}}{\omega} = C_P(\lambda, \beta) \frac{1}{2} \rho A_v v_r^2 r. \tag{17}$$

where ω is the turbine rotor rotation speed in rad/sec.

1.2.4 *Determination of C_P as function of λ and blade pitch angle β*

Simplification can be applied to evaluate the relationship of the power coefficient C_P w.r.t. tip-speed ratio λ to plot C_P–λ curves as a function of pitch angle β. Figure 7 shows C_P–λ characteristic curves for two different horizontal axis wind turbines with β as a variable parameter.

If turbine C_P–λ curves as shown in Fig. 7 are available through direct experimental measurement or analytical function calculations, a data field

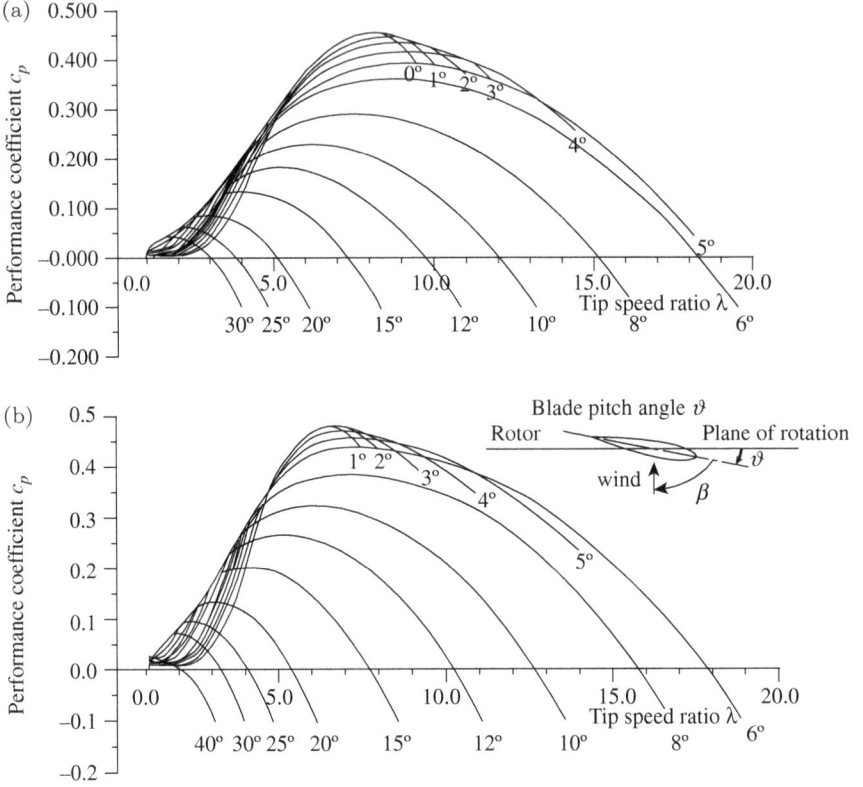

Fig. 7. C_p–λ curves at different blade pitch angles (a) a two-blade turbine with 12.5 m rotor diameter, 20 kW nominal power and (b) a three-blade turbine with 60 m rotor diameter 1.2 MW nominal output power.[4]

can be created by reading off the various values or entering them directly. These then form the basis for power coefficient computation in system simulations. When enough data are available, linear interpolation can be used to arrive at intermediate values.

A model defining the rotor power coefficient C_P as a function of tip-speed ratio and pitch angle β, $C_p(\lambda, \beta)$,[4] as shown in Fig. 8 is:

$$c_p(\lambda, \beta) = c_1 \left(\frac{c_2}{\lambda_i} - c_3 \beta - c_4 \right) e^{\frac{-c_5}{\lambda_i}} + c_6 \lambda,$$

with

$$\frac{1}{\lambda_i} = \frac{1}{\lambda_i + 0.08\beta} - \frac{0.0035}{\beta^3 + 1}. \tag{18}$$

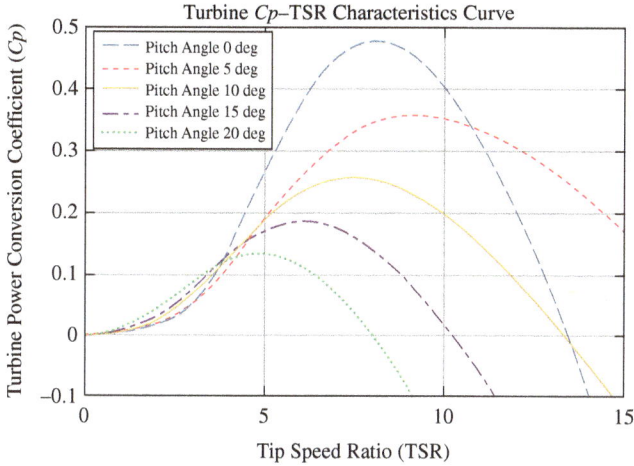

Fig. 8.   Approximated C_P–λ characteristic curves.[4]

The coefficients c_1 to c_6 are: $c_1 = 0.5176$, $c_2 = 116$, $c_3 = 0.4$, $c_4 = 5$, $c_5 = 21$ and $c_6 = 0.0068$[4]. The C_p–λ characteristics, for different values of the pitch angle β, are illustrated in Fig. 8. The maximum value of c_p ($c_{\mathrm{pmax}} = 0.48$) is achieved for $\beta = 0$ degree and for $\lambda = 8.1$. This particular value of λ is defined as the nominal value (λ_{nom}).

1.3   *Wind turbine power control*

The ambient wind in any location is variable for both speed and direction. In addition, a turbine is subject to turbulent– wind gusts of a transient and unpredictable nature.

1.3.1   *Turbine power control at different wind speeds*

The Wind Turbine Power Curve is as shown in Fig. 9,[7] which varies according to the following wind speed ranges:

- Cut-in speed $v_{\mathrm{cut-in}} = 2.5$–$4.5\,\mathrm{m/s}$
- Design wind speed $v_{\mathrm{d}} = 6$–$10\,\mathrm{m/s}$
- Nominal wind speed $v_{\mathrm{n}} = 10$–$16\,\mathrm{m/s}$
- Cut-out wind speed $v_{\mathrm{cut-out}} = 20$–$30\,\mathrm{m/s}$
- Survival wind speed $v_{\mathrm{life}} = 50$–$70\,\mathrm{m/s}$

Power control at three different wind speed ranges is described in the following.

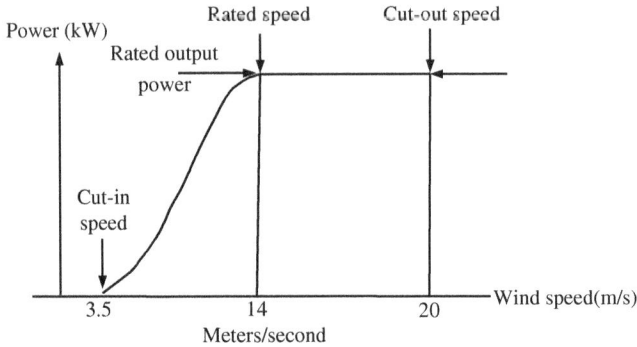

Fig. 9.   Wind turbine power curve.[7]

1.3.1.1   Cut-in speed

At very low wind speeds, there is insufficient torque exerted by the wind on the turbine blades to make them rotate. The speed at which the turbine first starts to rotate and generate power is called the **cut-in speed** and is typically between 3 and 4 meters per second. When the wind speed increases further, the wind turbine will rotate faster and generate electrical power.

1.3.1.2   Rated output power and rated output wind speed

As the wind speed rises above the cut-in speed, the level of electrical output power rises rapidly as shown. However, typically somewhere between 12 and 17 meters per second, the power output reaches the limit that the electrical generator is capable of. This limit to the generator output is called the **rated power output** and the wind speed at which it is reached is called the **rated output wind speed.** At higher wind speeds, the design of the turbine is arranged to limit the power to this maximum level and there is no further rise in the output power. How this is done varies from design to design, but typically with large turbines it is done by adjusting the blade angles so as to keep the power at a constant level.

1.3.1.3   Cut-out speed

For reasons of mechanical safety and to avoid structural damage, a wind machine can only operate up to a maximum design value of rotational speed, which is associated with the wind speed. As the speed increases above the rated output wind speed, the forces on the turbine structure continue to rise, and, at some point, there is a risk of damage to the rotor. As a result, a braking system is employed to bring the rotor to a standstill. This is called the **cut-out speed** and is usually around 25 meters per second.

1.3.2 *Fixed rotation speed and variable rotation speed*

Wind turbines can be broadly classified into two types — fixed speed and variable speed — which employ different control methods.[8]

(a) Operating a turbine at variable rotational speed following the wind, or
(b) Regulating the speed of rotation to create a fixed speed or a choice of two different fixed speeds of rotation.

For either operation, the turbine must be capable of being completely stalled into total immobility at some predetermined safe maximum operating speed.

Constant speed operations: This may be realized by

1. Varying the pitch angle of the propeller blades as the wind speed varies, using pitch angle control.
2. The use of a propeller of fixed pitch angle, but the propeller surfaces are designed to introduce stall over a wide range of wind speeds.

When the turbine rotational speed is constant, a coupled A.C. generator will operate at a fixed frequency which can be synchronized to the frequency of the grid to which it is connected. It is not necessary to use any form of electronic frequency changer as a decoupled device as is required in variable speed system. This kind of system is simple and cheap.

Variable speed operations: This enables the turbine speed to vary as the wind speed changes to keep TSR close to the optimal value. This is more efficient than the fixed speed control by about 20% to 30%, but is more expensive. The additional expense of variable speed control arises from the need for a controlled variable pitch propeller and the need for a power electronic converter.

1.3.3 *Three methods of blade control to prevent turbine overloading*

These are Passive Stall Control, Active Stall Control and Pitch Angle Regulation, which are also called "aerodynamic controls". The main objective of aerodynamic control is to protect the generator, the turbine and other mechanical parts from overloading during a high wind speed condition. The three types of aerodynamic controls are shown in Fig. 10.

Passive stall control (Fig. 10(a)): this is generally applied to fixed speed wind turbines for the vast majority of wind turbines in the class up to 1 MW.

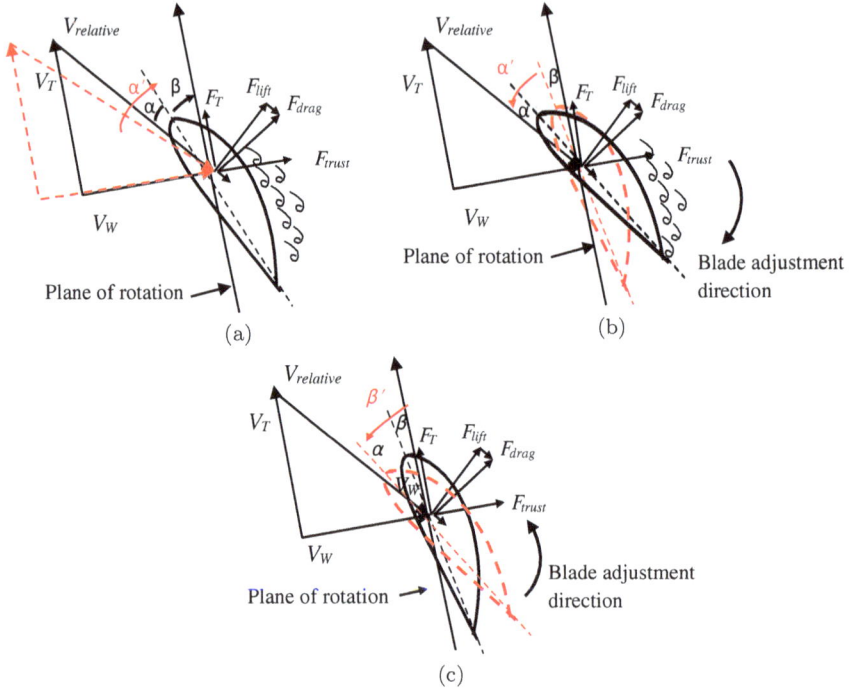

Fig. 10. Three different methods of blade control (a) passive stall control, (b) active stall control and (c) pitch angle control.

These are without a blade positioning mechanism and are fitted with asynchronous generators.

The blades are bolted to the hub at a fixed angle. This control depends only on wind speed. According to the aerodynamics of the airfoil, as the wind speed increases, the relative wind speed also increases resulting in an increase of the angle of attack α. If α is below its critical limit, the wind-flow above the blade is still a streamline. When wind speeds exceed nominal levels, the angle of attack, α, increases. This streamline turns to turbulence and becomes complete turbulence if α is above its critical limit. This results in a decrease in lift coefficient $c_a = f(\alpha)$, hence a lift force and an increase in $c_w = f(\alpha)$, hence a drag force. As a result, the torque-creating tangential force does not exceed the nominal values, even though the turbine is under full load and the wind speed climbs beyond the nominal range. The power coefficient is low despite the high energy available. In this method of control, the pitch angle remains constant. the rotor and blade rotating speeds are dependent on the wind speed when it is within the nominal level.

The disadvantage: there is a low possibility to influence operations.

Active stall control (Fig. 10(b)): this is similar to passive stall control, but instead of having a fixed blade angle, there are mechanical rotational parts to turn the rotor blades in the direction of the plane of rotation (opposite to the pitch angle control). This increases the angle of attack α, moving the turbine into the stall range, thus reducing the power drawn from the airstream. This helps to avoid any overshoot of the rated power of the machine due to the appearance of a wind gust. In general, a blade pitch-adjustment range of a few degrees is sufficient to protect the machine from overload. The method is applied to fixed-speed wind turbines which can operate only at one or two ideal wind speeds.

Pitch angle control (Fig. 10(c)): adjusting the pitch angle of the blades provides an effective means of regulating or limiting overloading the blades in high wind speeds or storm conditions. Over a range of wind speeds from cut-in wind speed to rated wind speed, the pitch angle β is adjusted to be zero or a small negative-angle to increase the power coefficient. For a wind speed from the rated value to the cut-off wind speed, the pitch angle is steadily increased to reduce the aerodynamic force of the wind. The direction of increasing blade pitch angle β is opposite to the direction of increasing angle of attack α as shown in the Fig. 10. Until reaching fully feathered mode, the edge of the blade turns against the wind. Therefore the stress on the turbine is reduced and the power coefficient C_p is reduced significantly. To achieve pitch angle control, hydraulic or electrical servo motors are used to adjust the pitch angle of the blades. To date, pitch angle control is applied to variable speed wind turbines with variable speed control of the generators to achieve the maximum possible power generation at any wind speeds.

1.3.4   *Wind turbine yaw control*

The wind at a given site is subject to rapid and frequent changes of direction. However, to maintain efficient operation the turbine propeller plane must be maintained perpendicular to the wind direction. This requires that the turbine assembly be free to rotate about a vertical axis — a phenomenon that aeronautical engineers call the "yaw" effect.[4] With good bearings, a machine can be pivoted to swivel under the influence of a vane so that

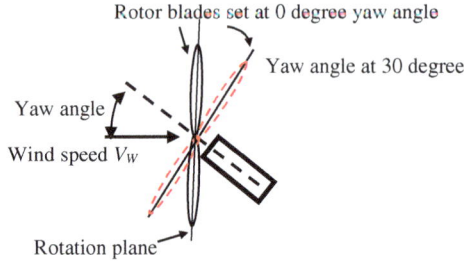

Fig. 11.   Wind turbine Yaw effect.

the plane of rotation can be changed according to the direction of wind pressure.

The horizontal axis of a wind turbine must always follow the direction of the wind. The orientation of the rotor blades must be chosen so that the rotor blades face the wind at the optimal angle. The position of a rotor can be upwind or downwind. Most large turbines use upwind. A simple wind vane, as might be used on a small machine, moves the rotor of an upwind turbine always to a position perpendicular to the wind. Turning wind turbines out of the wind reduces the effective flow cross-section of the rotor, causing a drastic drop in the performance coefficient.

As shown in Fig. 11, the yaw angle can be set to 0 degrees. Hence, the effective sweep area of blade $A = A_{\text{norm}}$, the maximum. If the yaw angle $=$ $30°$, the effective sweep area of blade $A = A_{\text{norm}} \sin(30°)$, the extracted power is reduced.

The wind turbine is said to have a yaw error if the rotor is not perpendicular to the wind. A yaw error implies that a lower share of the energy in the wind will be running through the rotor area. Yaw angle is the angle between the rotation axis and the wind direction.

That part of the rotor which is closest to the source direction of the wind will be subject to a larger force (bending torque) than the rest of the rotor. On the one hand, this means that the rotor will have a tendency to yaw against the wind automatically. On the other hand, it means that the blades will be bending back and forth in a flapwise direction for each turn of the rotor.

Wind turbines which are running with a yaw error are therefore subject to a larger fatigue load than wind turbines which are yawed in a perpendicular direction against the wind direction. In large modern wind turbines, a weather vane monitors the wind direction, and an electric yaw drive is used to swivel the propeller plane broadside into the wind.

1.3.5 *Gyroscopic forces and vibration*

Yawing rotation about a vertical axis while the rotor blades are rotating about the horizontal axis encounters strong gyroscopic forces. These forces have to be transmitted through bearings and propeller shafts, causing high stresses and vibrations. For this reason, the propeller blades of large machines are made of a lightweight material such as composite plastic like fiberglass rather than metal.

The action of rotation of the blades results in periodic vibrations. With a downwind-designed machine, each rotating blade passes through the wind shadow of the tower once per rotation. This results in a sudden reduction of air pressure on each blade followed by a sudden increase of air pressure as it emerges from the shadow of the tower. This results in the application of a bending moment on each blade at its root or hub joint in alternate direction. Continual flexing of the propeller blades at every rotation produces fatigue stresses in the materials.

1.4 *Shaft torque and power*

The basic external operating characteristic for any rotational machine is the shaft torque T versus the shaft rotational speed ω. A separate $T - \omega$ characteristic is obtained for each different value of wind speed at the turbine rotor. Typical forms of this characteristic are shown in Fig. 12 for a particular turbine. The locus of the maximum torque T_m is a quadratic function of the speed in the form of $T_m = K_1\omega^2$, where K_1 is constant and ω is the rotational speed in radians per sec.

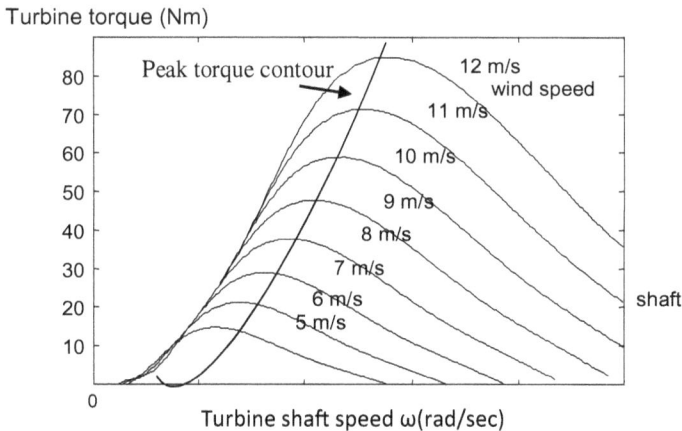

Fig. 12. Wind turbine Torque-Speed. curve.[10]

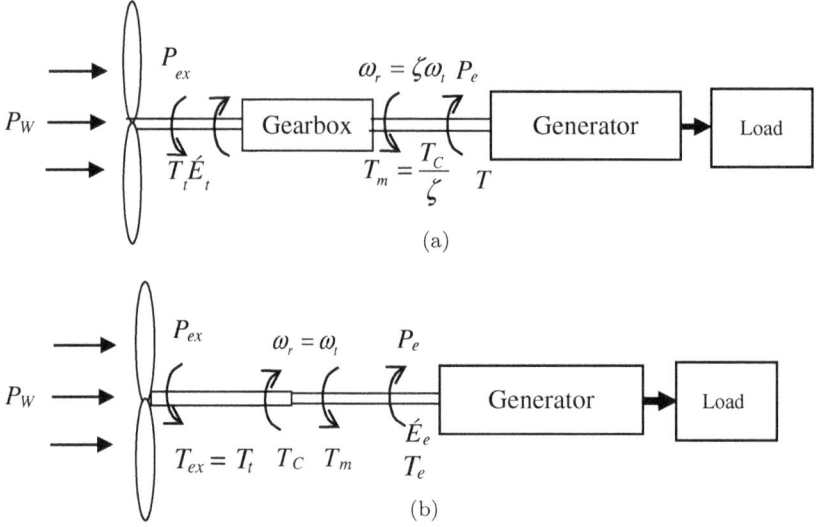

Fig. 13. Torque, speed and power at various stages of a wind power generation system. (a) Geared machine (b) Gearless machine.

The wind turbine is usually coupled to a generator via a gearbox as in Fig. 13(a) to step up the generator shaft speed, or directly coupled to the generator as in Fig. 13(b). For this reason, the generator is usually mounted at the top of the supporting tower along with the gearbox. Electric cables run down the tower to connect the generator to its electric load on the ground below.

The torque, speed and power of a rotating shaft are linked by the relationship $P = T\omega$. At zero speed, the shaft power is also zero. The shaft power is also zero when the torque is zero. Maximum shaft power P_m increases with the increase of wind speed. The locus of the maximum power points is a cubic function with the form $P = K_2\omega^3$ where K_2 is constant.

Geared machine (Fig. 13(a)): The gear box normally presents in a system using a high-speed asynchronous generator, and in this case the drive shaft is relatively long; there will also be a second, usually shorter, drive shaft connecting the gearbox to the generator.

The gearbox may add fatigue torque loading, related reliability issues and maintenance costs.

Using a multi-pole synchronous generator, the gearbox may be absent and in this case the drive shaft is relatively short.

Gearless machine (Fig. 13(b)): The rotor shaft is attached directly to the generator, which spins at the same speed as the blades. The diameter of the generator's rotor is increased so that it can contain more magnets to create the required frequency and power. Compared to gear-based generators, the advantages include increased efficiency, reduced noise, longer lifetime, high torque at low rpm, faster and precise positioning,

Experts from the Technical University of Denmark[8] estimate that a geared generator with permanent magnets may use 25 kg/MW of the rare-earth element (Neodymium),[9] while a gearless may use 250 kg/MW. China produces more than 95% of the world's rare-earth elements, while permanent magnet wind turbines only account for about 5% of the market outside of China, their market share inside China is estimated at 25% or higher.

Torsional Stress:[2,6] To clarify terms, we have referred to torque as the total twisting force produced by, or acting on, a system component such as a turbine or generator. Angular twist or displacement produced by torque acting on the elasticity of a shaft is referred to as torsion, and shear stress distributed over the cross-section of a shaft is termed torsional stress. Torque is the total twisting effort of torsional stress integrated over a shaft cross-section.

Wind turbine blades and generators are connected to the shaft at different points over its length and have different moments of inertia. Shaft systems can clearly exhibit torsional oscillations, and have resonance frequencies that can amplify torsions. Oscillations are mainly damped over time by wind turbine mechanical losses, but there is little additional frictional damping unless it is introduced deliberately.

A simplified analysis on the shaft torsion stress, as illustrated in Fig. 14, is as follows.

The rotational forces on the turbine propeller impose a torque on the gearbox turbine side which is $T_t = P_{ex}/\omega_t$ Similarly the torque on the gearbox output shaft to the generator is $T_g = P_g/\omega_g$.

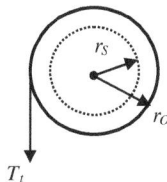

Fig. 14.   Torsional stress on turbine shaft.[2,6]

For a solid cylindrical shaft, subject to a total torque $T = T_t - T_g$, the torsional shear stress f_S acting on the shaft at any arbitrary radius r_S is given by

$$f_S = \frac{T \cdot r_S}{J} N/m^2, \tag{19}$$

where J is the polar (area) moment of inertia, having the dimension m^4. For a solid cylindrical shaft of radius r_O, the polar moment of inertia J is given by

$$J = \frac{\pi \cdot r_O^4}{2} m^4. \tag{20}$$

Combining the above two formulae gives an expression for the shear stress at the surface of the solid-cylindrical shaft of radius r_O of

$$f_S = \frac{2T}{\pi r_O^3} N/m^2. \tag{21}$$

1.5 Dynamics of drive train and two mass model[12]

The main parts of a wind turbine generator drive train are the turbine itself, a drive shaft, a gearbox if present, and the generator itself. In general, about 90% of the moment of inertia of the whole drive train is formed by the hub and blades together, while the generator rotor typically contributes 6–8%.[11] There are also gears, brake disc, clutches and shafts, but these count little in terms of moment of inertia. Hence, they are approximated as having zero mass when considering the transmission system behavior.

For a wind power generation system with or without gears, the extracted torque which is also the turbine driving torque T_T acts on the mechanical drive train from one end and the load torque T_e of the genera-tor acts on it from the other end. Between the two main components, there exists an angular speed coupling via the mechanical element connecting them. The rigidity and damping characteristics of the individual compo-nents, and any play in the clutch and gear, exert a decisive influence on the behavior of the transmission. Thus, when analyzing the turbine drive train the principle for a rotational system consisting of a disk with moment of inertia J mounted on a shaft fixed at one end is applied. We assume the viscous friction coefficient (damping) is D and the shaft torsion spring constant (stiffness) is K. So the torque on the disk can be calculated from the free-body diagram of the disk[12, 13] as

$$T(t) = J\frac{d^2\theta(t)}{dt} + D\frac{d\theta(t)}{dt} + K\theta(t), \tag{22}$$

where $\theta(t)$ is the angular value of rotation and $T(t)$ is the disk driving torque.

As with the wind turbine, the driving torque T_t is derived from an extractable mechanical power equal to the product of torque and angular velocity, and can be derived from

$$T_t = \frac{P_{ex}}{\omega_t} = \frac{1}{2}\frac{C_p(\lambda, \beta)\rho A V_w^3}{\omega_t}. \tag{23}$$

The torque at generator coupling T_C can be represented by the simplified form

$$T_C = k_{TS}(\theta_t - \theta_g) + k_D\frac{d(\theta_t - \theta_g)}{dt}. \tag{24}$$

Or $T_C = T$ (due to torsional elastic component of drive train) $+T_D$ (due to damping component). At a stationary situation $T_D = 0$, so the coupling torque reduces to the torsional torque. So the torsional stiffness coefficient can be determined by

$$k_{TS} = \frac{T_C}{\theta_t - \theta_g} = \frac{T_C}{\Delta\theta}.$$

Thus, the model on the turbine side is

$$J_T\frac{d\theta_t^2}{dt^2} = T_t - \left(k_{TS}(\theta_t - \theta_g) + k_D\frac{d(\theta_t - \theta_g)}{dt}\right). \tag{25}$$

On the generator side T_m is the drive torque which equals T_C when there is no gear and neglecting the transmission losses. Hence, the model for the two mass systems without a gearbox is:

$$\frac{d}{dt}(\theta_t - \theta_g) = \omega_t - \omega_g, \tag{26}$$

$$\frac{d\omega_t}{dt} = \frac{1}{J_T}(T_t - k_{TS}(\theta_t - \theta_g) - k_D(\omega_t - \omega_g)) \tag{27}$$

$$\frac{d\omega_t}{dt} = \frac{1}{J_g}(k_{TS}(\theta_t - \theta_g) + k_D(\omega_t - \omega_g) - T_e), \tag{28}$$

where,

ω_t, ω_g = wind turbine rotor and the generator speed [rad/sec],
θ_t, θ_g = angular positions of the turbine rotor and the generator rotor,
k_D and k_{TS} are the equivalent damping and stiffness of the wind turbine rotor.

With a gearbox and assuming its gear ratio is ζ in order to maintain the generator shaft speed within a desired speed range, the torque and shaft speed of the wind turbine referred to the generator side of the gearbox are given by:

$$T_m = \frac{T_C}{\zeta} \quad \text{and} \quad \omega_g = \zeta\omega_t,$$

where T_m becomes the driving torque of the generator and ω_r is the generator shaft speed. When the generator is loaded, the electric torque T_e acts on the generator shaft as a load, imposing the drive train with a torque in reverse rotation direction.

If we refer to the turbine side, the model equations are[12]

$$\frac{d}{dt}\omega_t = \frac{1}{J_T}[T_T - k_D(\omega_t - \omega_g) - k_{TS}(\theta_t - \theta_g)], \tag{29}$$

$$\frac{d}{dt}\omega_g = \frac{1}{\zeta J_g}[+k_D(\omega_t - \omega_g) + k_{TS}(\theta_t - \theta_g) - \zeta T_g], \tag{30}$$

$$\frac{d}{dt}(\theta_t - \theta_g) = \omega_t - \omega_g. \tag{31}$$

However, modeling of the above is complicated and often, in analysis of wind turbine generator system operations in electric power systems, the wind wheel-dynamics are neglected. The following provides only a simplified analysis.

$$\frac{T_m - T_e}{T_N} = J_R \frac{\omega_n}{T_N}\frac{d(\omega/\omega_n)}{dt}, \tag{32}$$

where J_R is the moment of inertia of all rotating masses, $T_m - T_e$ is the acceleration torque of the rotating components, T_N and ω_N are, respectively, the nominal torque and angular velocity.

1.6   *Efficiency consideration of wind powered electricity*

A flow diagram for a basic wind converter system is given in Fig. 15. The combination of turbine, gearbox and generator is sometimes called a "power

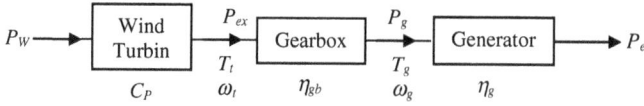

Fig. 15. Efficiency analysis.[6]

train". Efficiencies of various stages are given by

$$Turbine\ efficiency = P_{ex}/P_w = C_p,$$

$$Gearbox\ efficiency = P_g/P_{ex} = \eta_{gb},$$

$$Turbine\ efficiency = P_{out}/P_g = \eta_g, \tag{33}$$

The overall efficiency of the three-stage system of Fig. 15 from input to output, is

$$\eta = \frac{\text{electrical output power}}{\text{power available in the wind}} = \frac{P_e}{P_w} = \frac{P_{ew}}{P_w} \cdot \frac{P_g}{P_{ex}} \cdot \frac{P_e}{P_g}. \tag{34}$$

In terms of the individual stage efficiencies, from Eq. (33), the overall efficiency η can be written as

$$\eta = \frac{P_e}{P_w} = C_p \eta_{gb} \eta_g. \tag{35}$$

For small wind energy installation, up to a few kW output rate, the overall efficiency η is of the order *20%–25%*.

The electrical output power may be written as

$$P_e = C_P \cdot \eta_{gb} \cdot \eta_g \cdot P_w = C_P \cdot \eta_{gb} \cdot \eta_b \cdot \frac{1}{2}\rho A V^3. \tag{36}$$

For the three-blade propeller type of wind turbine,[2]

$C_P = $ *40–50%* for large machines (100 kW–3 MW)
$C_P = $ *20–40%* for small machines (1 kW–100 kW).

1.7 *Solidity factor*[6] (*SF*)

This is defined as the total blade area of the rotor divided by the swept area normal to the wind. Low solidity factor is usually more desirable. The turbine rotor consists of the blades and the nose cone. The blades are commonly made of carbon fiber composites and are designed to achieve the optimum aerodynamic shape. The number of blades varies from two to three. In general, the more blades there are, the slower the rotational speed.

Three blades, the "Danish concept", have the particular advantage that the polar moment of inertia with respect to yawing is constant, and is independent of the azimuthal position of the rotor. Two bladed designs have the advantage of reduced blade cost, but have the problem of instability of the blades in high wind. They require an adjustable axis for the hub (teetering hub) to avoid heavy shock to the turbine. Moreover, the two blade wind turbine requires high rotational speed for a given power, which results in increased noise. Considering the energy captured, the more blades there are, the smoother is the energy captured. However, more blades mean higher cost and higher moment of inertia.

Modern horizontal axis wind generators have one, two or three rotor blades. The lower the number of blades, the less the material used.

- Single-bladed rotors must have a counter-weight on the opposite side of the rotor; they do not have a smooth motion, hence there is high material stress.
- Two-bladed rotor requires fewer materials, and has a lower SF.
- Three-bladed rotors are most stable and provide an optically smoother operation and hence are visually integrated better into the landscape. Mechanical strain is also lower than for two bladed rotors.

2   Part 2: Wind Powered Generators and Systems

2.1   *Types of wind powered generation systems*

There are several different wind powered generation systems classified according to the generators used and the ways they interface to the grid. The following are the ones most widely used.

2.1.1   *Constant speed constant frequency (CSCF)*

The configuration of this system is a "Danish-concept"[14] shown in Fig. 16. It consists of a squirrel cage induction machine connected to the grid via a voltage controller. Both the voltage and frequency at the machine terminals depend on the shaft speed. But the grid supply frequency and voltage are fixed. Thus, only one particular speed of rotation is consistent with the system frequency.

Voltage regulation is required to prevent a large inrush current when connecting to the grid and there is mechanical stress between the supply grid and turbine. A step-wise controllable capacitor bank for reactive power compensation is inserted in parallel with the generator. This system

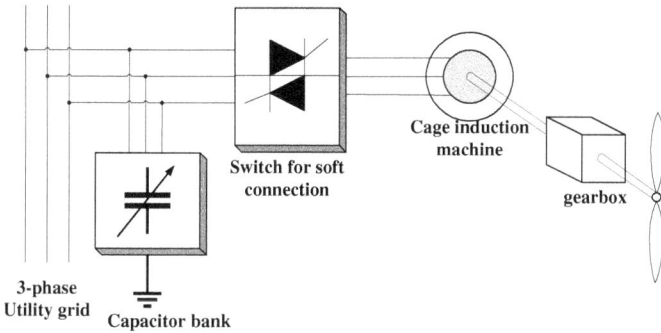

Fig. 16.   Constant speed wind turbine with a direct grid connected squirrel-cage induction generator.

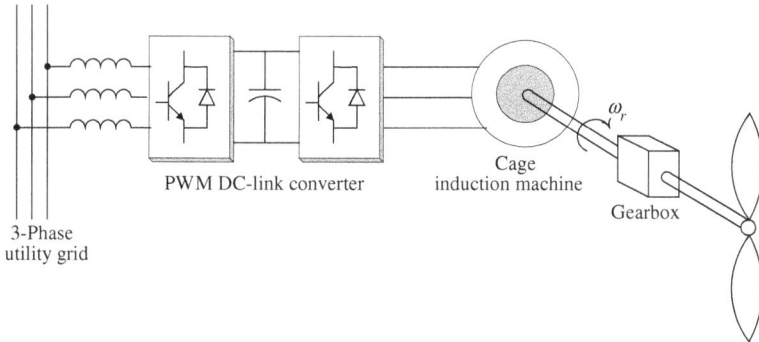

Fig. 17.   VSCF system using cage induction machine and AC-DC-AC converter.

is optimal at only one or two wind speeds. The power can be controlled only by blade pitch angle as described earlier.

2.1.2   *Variable speed constant frequency (VSCF)*

(a) A squirrel cage-rotor induction machine rotated by wind energy is shown in Fig. 17. Its stator is connected to the utility grid through a frequency changer consisting of a PWM DC-link converter.

The main advantage of this configuration lies in the simple and robust structure of the squirrel cage induction machine. The system can operate over a wide range of wind-speeds and permits a soft start-up. However, the converter rating needs to be approximately 125% of the machine rating as the reactive power required by the generator must be supplied by the converter.

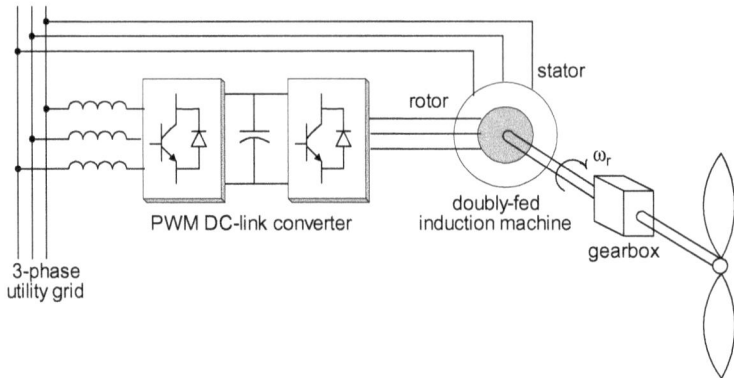

Fig. 18.   VSCF system using doubly-fed induction machine and AC-DC-AC converter.

(b) Doubly Fed Induction Generator (DFIG) consists of a wound-rotor induction generator having its rotor connected to the grid through a PWM DC-link converter. The machine stator is directly connected to the supply grid. A simplified block diagram of this system is shown in Fig. 18.

In such a configuration, the converter rating required is only the slip-fraction of the machine power rating due to a restricted operating speed range. Moreover, the real power is supplied from the machine stator terminals directly to the utility grid. On the rotor side, the AC-DC-AC converter controls the rotor voltage and current flow, hence the speed of the DFIG. It also adjusts the power factor at the stator terminals. The frequency of rotor current and voltage varies, depending on the turbine rotating speed.

(c) A synchronous generator driven by a wind turbine can also have its stator connected to the supply utility through a PWMDC-link converter as shown in Fig. 19. Similar to the squirrel cage induction machine-based CSVF, it can operate over a wide range of wind-speeds. The rotor magnetic field can be excited by a thyristor rectifier with its ac side connected to the generator-side converter or simply using a permanent magnet as the rotor.

2.2   Electric generation machines in wind energy systems

In most electrical power systems, the bulk of the power is supplied by rotating machine generators. These can be of many types, but those used for wind power generation are mainly the following:

- Director current (DC) machines,
- Alternating current (AC) synchronous generators and
- Induction generators.

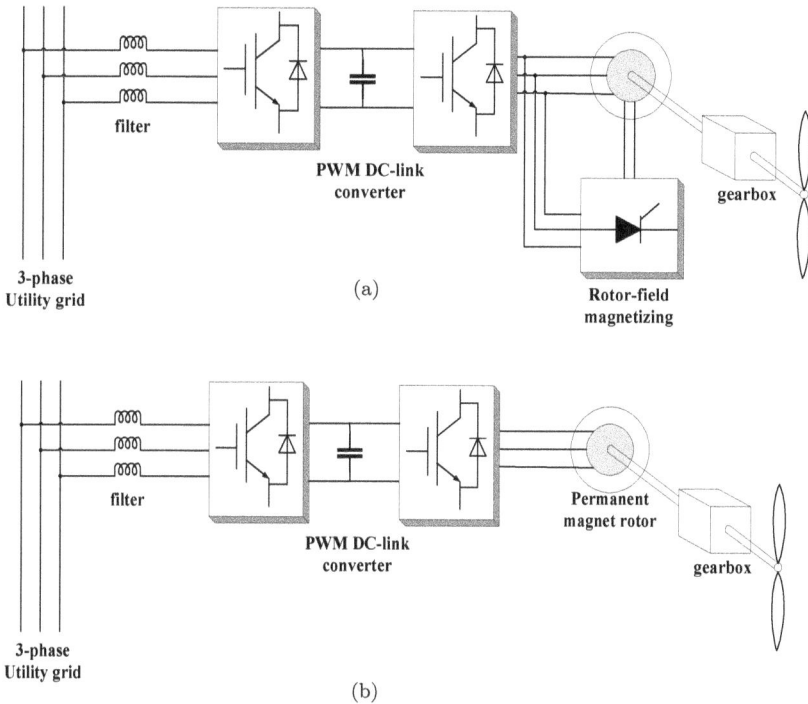

Fig. 19. VSCF synchronous generator system (a) with thyristor field excitation, (b) with permanent magnet rotor.[2,6]

2.2.1 *DC generators*

Small stand-alone wind generator systems were at one time fitted with shunt-wound DC generators. This form of self-excited generator is easy for speed control and was used extensively until the early 1980s, especially when the output power could be used in direct current form. The need for a machine with a commutator and brush gear resulted in low reliability and high maintenance costs.

Modern brushless DC machines use a permanent magnet as a rotor and the stator winding as armature called a permanent magnet DC (PMDC) machine. Using rare earth permanent magnets such as Neodymium or Samarium Cobalt, the stator field flux produced is very strong compared to the conventional brushed DC machine using wound coil stator excited by the DC current from the armature. The configuration of a PMDC generator essentially forms an inside-out conventional DC machine with its mechanical commutators replaced by an electronic commutation circuit. The advantages of a PMDC generator are:

- generally lighter than wound stator machines for a given power rating,
- better efficiencies because there are no field windings and field coil losses,
- resistant to the effects of possible dirt ingress, as the stator is provided with a permanent magnet pole system.

This brushless machine technology greatly improves machine reliability, but is restricted to machine ratings of less than a few hundred watts at most. As such, it is not suitable for the multi-megawatt machines installed in modern wind energy generation schemes.

2.2.2   AC generators

Most electricity supply systems are of three-phase alternating current form with fixed frequency and voltage, and use sinusoidal voltages and currents. Electrical power for modern transmission and distribution systems is invariably generated by three-phase, synchronous generators of many megawatts ratings. The largest of such machine sets in the U.K. are rated at 600 MW and are housed, with their fossil or nuclear fueled turbines, in large utility power plants or power stations.

Wind energy generators are several orders smaller in size, but are increasing in rating all the time.

Alternating current generators in wind power systems fall into two main categories:

(i) Induction generators, requiring alternating current excitation and generating a voltage at the same frequency as the excitation, with a magnitude dependent on the speed and direction of rotation.

(ii) Synchronous generators, requiring direct current winding excitation or permanent magnet excitation, that deliver a voltage and frequency which are both proportional to the rotation speed.

2.3   Three-phase induction generators

2.3.1   Operating principles and equivalent circuit

A three-phase induction machine, whether acting as a motor or a generator, has a double cylindrical structure. Figure 20[2, 6, 16] shows induction machines having either a cylindrical rotor containing distributed three-phase windings (wound rotor), or a cage rotor structure with copper bars as rotor conductors (cage rotor). The rotor is free to rotate at a speed, n_r (rpm), inside the cylindrical cavity of a stator structure built onto the stationary frame of the machine. The three-phase windings on the machine stator

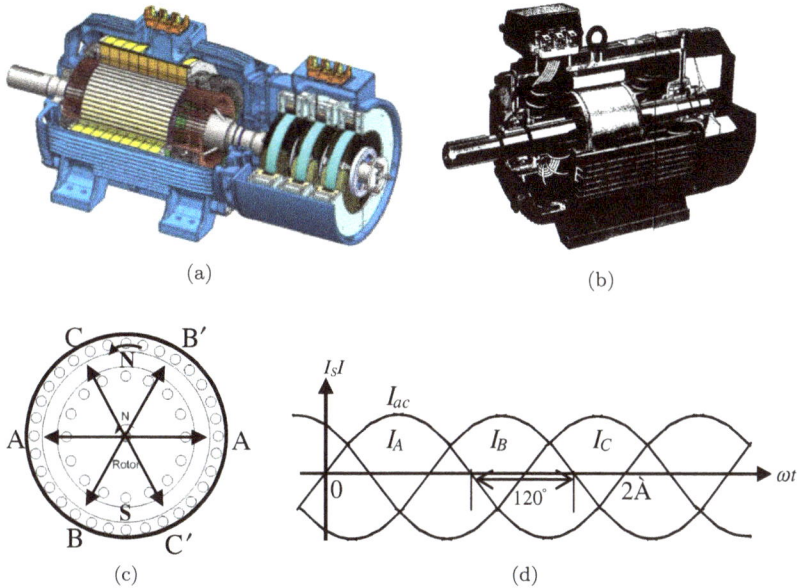

Fig. 20. (a) A wound rotor induction machine,[24] (b) a squirrel cage induction machine,[15] (c) rotating magnetic field in the air-gap and (d) three-phase sinusoidal current in the induction machine stator windings.[6]

are usually connected to a three-phase supply of constant frequency and voltage. These windings become the primary windings of the machine and are the excitation windings. Machine torque and electromagnetic power are developed in the secondary windings on the machine rotor. These windings are closed on themselves directly or through equal resistors. For wound rotor machines, an external circuit can be connected into the machine rotor secondary windings via slip-rings and brush-gear on the rotor.

When a set of balanced, three-phase sinusoidal voltages is applied to the three-phase stator windings, balanced three-phase currents flow in the stator windings. These produce a rotating magnetic field as though there were a pair of magnetic poles rotating around the air-gap. The speed of rotation of the field is called the synchronous speed, n_s (rpm), depending on the applied voltage frequency f_s (Hz), and the number of magnetic poles, defined by

$$n_s = \frac{2 \times 60 \times f_s}{poles} \text{rev/minute} \quad \omega_s = \frac{2 \times 2\pi f_s}{poles} \text{rad/sec.} \quad (37)$$

An induction machine operates by the principle of electromagnetic induction. The rotor conductors rotate in the direction of the stator rotating field at a velocity, n_r, that is lower or higher than the synchronous rotating velocity, n_s. The differential speed between the rotating stator flux and the actual rotor speed is called the slip speed, where $slip\ speed = n_S - n_r$.

The ratio of the slip speed to the synchronous speed is the most important variable in induction machine operation and is referred to as the "per-unit slip" s, given as

$$s = \frac{n_S - n_r}{n_S} = 1 - \frac{n_r}{n_S}. \tag{38}$$

An alternative form of this relationship is the expression for rotor speed,

$$n_r = (1 - s)n_S. \tag{39}$$

If the rotor operates at synchronous speed, then $n_r = n_s$ and slip $s = 0$. Because the rotor conductors cut the rotating flux field at the slip speed, the electro-magnetic forces (e.m.f.s), the rotor voltage induced in them are of slip-frequency and the rotor currents that flow in the rotor windings are also of slip-frequency. Thus, for the supply stator frequency f_s the corresponding rotor voltage and current frequency, i.e. the slip-frequency, f_r, is given by $f_r = sf_S$.

For an induction motor, the rotor rotating speed, n_r, is lower than the synchronous speed n_s, hence the per unit slip is positive and its magnitude is usually below 10%, i.e $0 \leq s \leq 0.1$. This implies that the frequency of the rotor electrical variables is $f_r \leq 0.1f_s$, which is 6 Hz at $f_s = 60$ Hz and 5 Hz at $f_s = 50$ Hz and in the same direction to that of the rotating magnetic field. For an induction generator with rotor speed varying from standstill to above the synchronous speed, the frequency of the rotor currents therefore varies from synchronous frequency f_s at the machine standstill to a few Hz higher than f_s.

Now the fields of the stator and rotor windings must be in synchronism with each other for steady torque to be generated. The stator flux rotates at the synchronous speed, n_s, relative to the stationary stator windings, while the rotor rotates at n_r. This requires that the rotor field must rotate with respect to the rotor at slip-speed sn_s which is positive when operating as a motor but negative as a generator. For the latter, the rotor field is in the reverse direction to its rotating direction. This is the key difference from when it is operating as a motor.

The two flux components on the two windings rotate in synchronism at n_S independently of the variable speed n_r of the machine shaft. The stator induced e.m.f. viz. $E_1 = 4.44f_sN_1\Phi_m$. Assuming a unity transformation

(a)

(b)

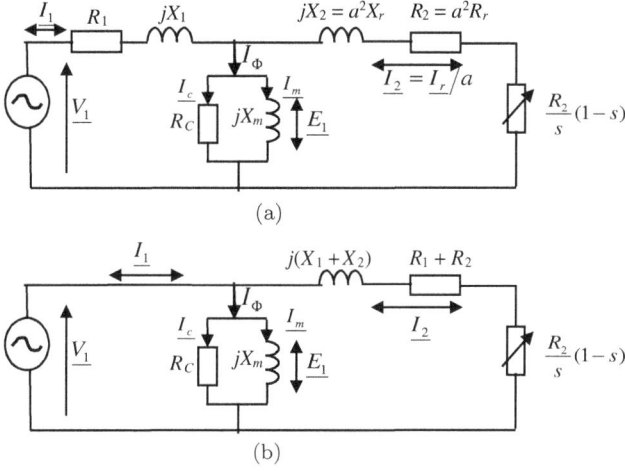

Fig. 21. (a) Simplified per-phase equivalent circuit with rotor current and Impedance referring to the stator side. (b) Simplified per-phase equivalent circuit.

ratio between stator and rotor windings $(N_1/N_2 = 1)$, the magnitude of the rotor e.m.f., E_2 in terms of E_1, is $E_2 = sE_1$, which is proportional to the per-unit slip s.

A per-phase equivalent circuit referred to the stator turns is shown in Fig. 21, from which the rotor current is (note the direction of current and voltage follows the motor convention):[15, 17]

$$I_r = \frac{E_2}{R_r + jsX_r} = \frac{sE_1}{R_r + jsX_r} = \frac{E_1}{(R_r/s) + jX_r}, \qquad (40)$$

where I_r is rotor side current, R_r is the rotor winding resistance in Ω, X_r is the rotor winding leakage reactance at synchronous frequency in Ω.

Considering the stator and rotor winding turns ratio a, the rotor side current and impedance referring to the stator side are:

$$I_2 = I_r/a, R_2 = a^2R_r, \quad X_2 = a^2X_r,$$

so the e.m.f. at the stator side is

$$E_1 = \frac{I_2R_2}{s} + jI_2X_2. \qquad (41)$$

The stator current, I_1, is the sum of machine magnetizing current and rotor current as follows:

$$I_1 = I_\Phi + I_2, \qquad (42)$$

where I_Φ is the magnetizing current, equal to the vector sum of iron core loss current I_c and excitation current I_m (as shown in Fig. 21(b)). The supply voltage can be expressed as

$$\underline{V}_1 = \underline{E}_1 + \underline{I}_1(R_1 + jX_1). \tag{43}$$

Figure 21(b) shows a simplified equivalent circuit, with rotor elements referred to the stator side, which neglects the voltage drop on the stator windings due to the stator current. So it is seen that the apparent current is:

$$\left|\underline{I}_2\right| = \frac{|\underline{V}_1|}{\sqrt{(R_1 + \frac{R_2}{s})^2 + (X_1 + X_2)^2}}, \tag{44}$$

Correspondingly, the electromagnetic torque is:

$$T_g = \frac{3\left|\underline{V}_1^2\right|}{2\pi f_s} \frac{\frac{R_2}{s}}{(R_1 + \frac{R_2}{s})^2 + (X_1 + X_2)^2}. \tag{45}$$

The positive sign corresponds to motor operation and the negative sign to generator operation as explained in the next sub-section.

2.3.2 Wind powered induction generator

For an induction generator, power passes into the machine through its shaft from the wind turbine. For a squirrel cage induction generator, power is injected into the external electrical system through the stator windings, while the air-gap flux is also excited from the grid through the same stator windings. For a wound rotor induction generator, power output to the grid can be from both stator and rotor sides, while excitation current is supplied from the grid to the rotor through power converters.

In this setting, in order for electrical generation to occur the rotational speed n_r of the rotor must exceed the stator synchronous speed n_s. At the synchronous speed $n_r = n_s$, the slip becomes zero, R_2/s becomes infinite and I_2 is zero, so the machine is "floating" on the bus. The only current is the exciting current supplying the rotating magnetic field and the iron losses. The input of mechanical power from the wind turbine results in increasing the rotor speed above synchronous speed, so $n_r > n_s$ so the slip-speed, s is now negative, and so is the slip frequency sf_s.

$$n_S - n_r < 0$$

Since the rotor conductors cut the rotating flux field at the negative slip-speed, the electro-magnetic forces (e.m.f.s) induced in them are of negative slip frequency and the rotor currents that flow are also of negative slip frequency, hence the rotor flux must rotate backwards.

The interaction of the magnetic flux of the stator and the magnetic flux of the rotor produces a counter-torque in opposition to the driving torque of the turbine. As the speed of the turbine is increased, the increase in electrical power supplied to the distribution system by the induction generator causes an increase in counter-torque. In this case, the torque given in equation (45) becomes negative by motor convention as

$$T_g = \frac{3}{2\pi f_s} \frac{|V_1^2|}{(R_1 - \frac{R_2}{|s|})^2 + (X_1 + X_2)^2}.$$ (46)

2.3.2.1   Speed-torque characteristics and pushover torque

This analysis applies to both generator and motor operation. According to the equivalent circuit in Fig. 21(b), the stator winding resistance R_1 is considered negligible and X_1 and X_2 are grouped as one, X_σ, the torque equation can be further simplified, so we have:

$$
\begin{aligned}
T_g &= \frac{3}{2\pi f_S} \cdot \frac{R_2}{s} \cdot \frac{V_1^2}{\left| \frac{R_2}{s} + jX_\sigma \right|} \\
&= \frac{3}{2\pi f_S} \frac{R_2}{s} \cdot \frac{V_1^2}{\frac{R_2^2}{s^2} + X_\sigma^2} = \frac{3V_1^2}{2\pi f_S X_\sigma} \cdot \cdot \frac{2}{\frac{R_2}{s X_\sigma} + \frac{s \cdot X_\sigma}{R_2}}.
\end{aligned}
$$ (47)

Set

$$T_B = \frac{3V_1^2}{2\pi f_S} \frac{2}{X_\sigma} \quad \text{and} \quad \frac{R_2}{X_\sigma} = s_B,$$

and the above torque equation becomes:

$$T_g = T_B \cdot \frac{2}{\frac{s}{s_B} + \frac{s_B}{s}}.$$ (48)

When $s = s_B$, the right-hand term becomes unity and the torque reaches the maximum. This is the so-called **breakdown torque** in motor operation. For generator operation, it is called the **pushover torque**, $s_B \approx (5 \text{ to } 10)s_n$. Figure 22 shows plots of torque and speed characteristics of an induction machine. Note the smooth transition from motoring to generating operations and the breakdown torque when operating as a motor. When in generator operation, the counter-torque increases with increasing

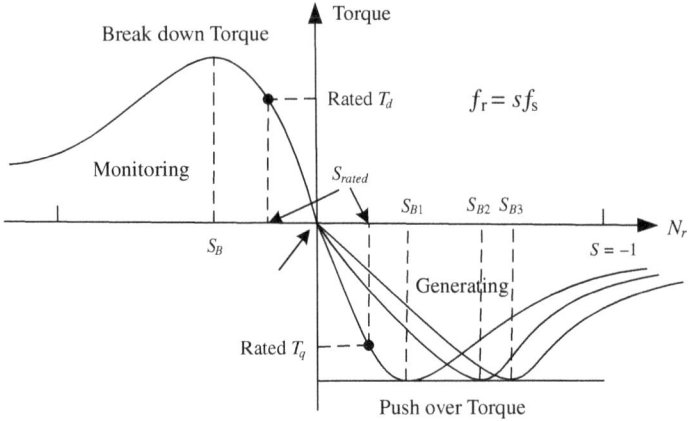

Fig. 22.   Torque/speed characteristics for both motoring and generating operation of a three-phase[17] generator. induction machine.

speed until it attains the maximum **pushover torque**, which is in contrast to **breakdown torque** when operating as a motor.

According to the above torque development, the maximum torque of an asynchronous machine is determined by the relation of ohmic to leakage components, particularly in the rotor windings. The size of the machine physically dictates the limit of slip. Large machines are highly limited in elasticity in the event of torque fluctuation. Generators having higher number of poles (6, 8 poles) show much more flexibility than two-pole machines. In machines with inflexible speeds of rotation, turbine torque fluctuations place heavy loads on the drive train because of almost rigid coupling between grid and generator. The rotating blades of the turbine are subject to considerable airstrip turbulence, especially near the tower. Even when the blades can react swiftly to limit turbine performance, the effect of tower shadow fluctuations cannot be ruled out entirely. For small slip machines, small speed variations can result in large variations in power.

Raising design slip levels can make the generator more flexible, hence offering more protection to the drive train. High slip values can be obtained by (1) high rotor resistance and (2) low total leakage reactance, hence lower rotor inductance.

In addition to setting higher slip values by adding more resistors in the rotor circuit, slip-ring rotor machines can be altered over a relatively wide spectrum using supplementary resistors or power converters in the rotor circuit to adapt to output power variations dynamically.

2.3.2.2 Power calculation

The available mechanical input power at the clutch reduced by losses due to mechanical friction and air resistance, P_f, is

$$P_{mech} - P_f = T_m \omega_r = T_g \omega_s (1 + |s|), \tag{49}$$

where T_m is the turbine driving torque, the developed electromagnetic torque T_g of the generator is considered balanced at the steady state with T_m.

This leads to the power through air-gap

$$P_{ag} = P_{mech} - P_f = 3 |I'_2|^2 \frac{R_2}{s} (1 + |s|). \tag{50}$$

According to motor convention this power is negative, i.e. power flows from the turbine rotor to the generator stator. According to the simplified equivalent circuit in Fig. 7, considering the copper losses in the stator and rotor,

$$P_{gloss} = P_{rotor} + P_{stator} = 3 |I'_2|^2 (R_1 + R_2), \tag{51}$$

I'_2 is the rotor current referred to the stator side and is evaluated by

$$I'_2 = \frac{V_1}{(R_1 + \frac{R_2}{s}) + j(X_1 + X_2)}. \tag{52}$$

Note the real element of I'_2 is negative due to the use of the motor convention. Consider that the rotor magnetizing current is

$$I_\phi = \frac{V_1}{R_C} + \frac{V_1}{+jX_m} = I_C - jI_m. \tag{53}$$

So the stator current to the grid is

$$I_1 = I'_2 + I_\phi = \hat{I}_1 (\cos\phi + j \sin\phi), \tag{54}$$

where φ is the phase angle between I_1 and V_1, and I_1 is the leading power factor. So the output power to the grid is

$$P_{out} = 3 \times V_1 \times I_1 \cos\varphi, \quad \text{and} \quad \text{efficiency} = \frac{P_{out}}{P_{mech}}. \tag{55}$$

Figure 23 shows the phasor diagram for an induction generator according to the equivalent circuit diagram of Fig. 21(a). Stator current phasor I_1

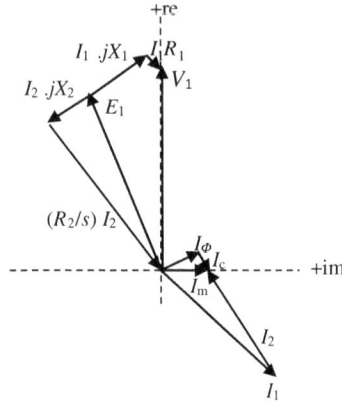

Fig. 23.   Phasor diagram for generator operation according to the equivalent circuit diagram in Fig. 21(a).[4]

is the sum of phasor $\underline{I_2}$ and current phasor in the magnetizing branch $\underline{I_m}$. The sum of $\underline{E_1}$ and voltage drops on stator resistance and reactance equals phasor voltage $\underline{V_1}$.

2.4   *Synchronous generator*[15, 18]

Synchronous generators are the principal sources of electrical power throughout the world, and range in size from a fraction of a kVA to a thousand MVA. Generators suitable for wind power turbines are in the range from a few kW to a few MW.

2.4.1   *Principles and equivalent circuit*

Two terms commonly used to describe the windings on a synchronous generator are field windings and armature windings. In general, the term field windings applies to the windings that produce magnetic field in the machine and the armature windings apply to the windings where the main voltage is induced and power is generated.

A three-phase synchronous machine is depicted in Fig. 24(a). Its stator consists of iron cores with three groups of coils wound around them. These identical coils, also named three-phase windings, have 120° phase shift between each of them and are connected directly to the three-phase power supply. Coils wound on the rotor core of a three-phase synchronous generator are energized by a DC voltage source. This sets up a field of magnetic flux which cuts the armature conductors as it rotates while driven by

(a) Synchronous generator*

(b) Salient pole rotor

(c) Cylindrical rotor

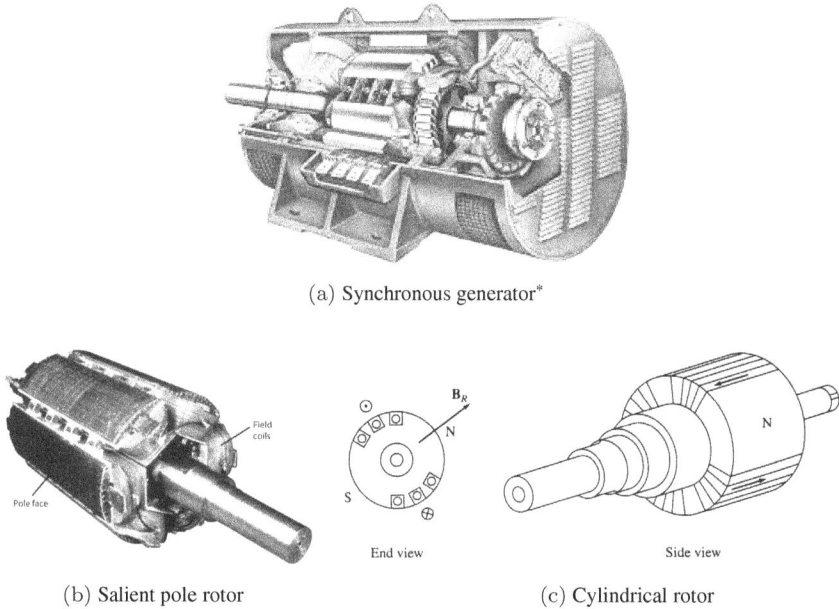

Fig. 24.   Synchronous generator and Two basic forms of rotor construction.[14]
Source: *https://www.avk-seg.co.uk/ (permission of AvK).

the prime mover — the wind turbine. The individual coils are connected in a manner that provides alternate north and south poles. Relative motion between the rotating field flux and the armature conductors causes e.m.f.s, i.e. voltages in the stator coils.There are two basic forms of rotor construction, (a) salient pole rotor and (b) non-salient or cylindrical pole rotor as shown in Fig. 24(b). Most wind generators use salient pole rotors or some use permanent magnet as rotor poles.

The rotor winding needs to be supplied with a DC current to establish the magnetic field. Since the rotor is rotating, the conventional method supplies the dc power from an external dc source to the rotor by means of slip rings and brushes. This is now replaced by other excitation methods, and one of them is called the Brushless Exciter, as shown in Fig. 25. A small ac generator with the field circuit is mounted on the stator and its armature circuit is mounted on the rotor shaft. The output of the armature is rectified to direct current by a three-phase rectifier also mounted on the shaft of the rotor and then fed to the main dc field circuit. By controlling the small dc field current of the exciter generator, it is possible to adjust

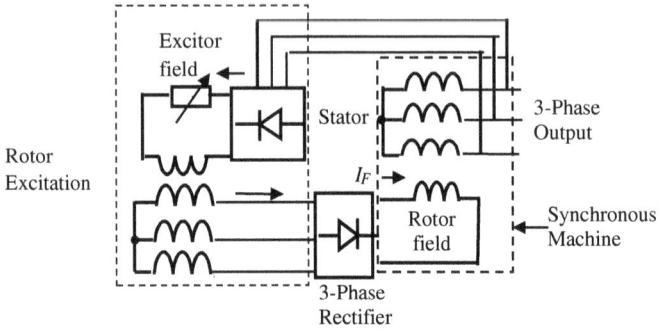

Fig. 25.   Synchronous generator field excitation using brushless exciter.[18]

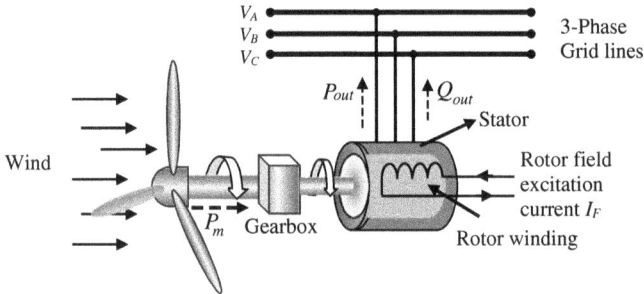

Fig. 26.   General form of a three-phase Wind-powered synchronous generator.

the field current of the main machine. Since there is no mechanical contact between stator and rotor, it requires less maintenance than using slip rings and brushes. Figure 26 illustrates the general form of a wind-powered synchronous generator.

Detailed description of the synchronous machine construction and performance are provided in many books on electric machines and are not included in the present text.

2.4.2   Power and torque of a synchronous generator

2.4.2.1   Speed of rotation and power balance

With a wind turbine as the prime mover, power flows into the machine characterized by the mechanical torque T_m on the rotor shaft and the speed of rotation ω_m (in S.I. units)[6,18] as

$$P_{in} = T_m \omega_m. \tag{56}$$

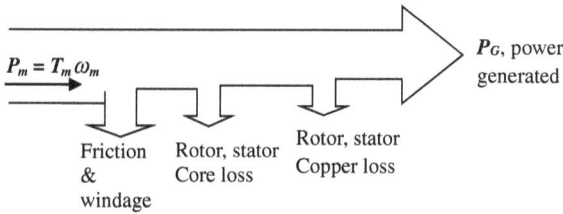

Fig. 27. Power flow diagram.

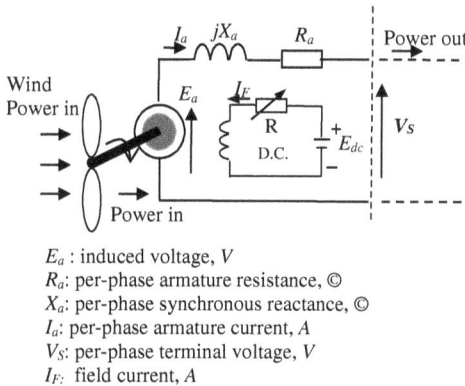

E_a : induced voltage, V
R_a: per-phase armature resistance, ©
X_a: per-phase synchronous reactance, ©
I_a: per-phase armature current, A
V_S: per-phase terminal voltage, V
I_F: field current, A

Fig. 28. Equivalent circuit diagram of a wind turbine synchronous generator.

The power, P_{in}, is in watts when the shaft torque T_m is in Newton meters and the shaft rotational speed ω_m is in radians/second.

All of the real power, P_G, generated by the machine comes in through the shaft from the turbine. Not all mechanical power going into a generator becomes electric power out of the machine. Losses include mechanical and stator and rotor. A power flow diagram for an alternator is shown in Fig. 27.

A synchronous generator connected to a load or a power supply system can be represented by the single phase equivalent circuit shown in Fig. 28. The generator frequency is determined by the rotor's rotational speed and the number of poles as

$$f = \frac{P_k \cdot N_S}{120}, \tag{57}$$

where f is the frequency in hertz (Hz), P_k is the number of rotor magnetic poles and N_S represents the shaft synchronous speed in rev/min (rpm). P_k is a design constant so that the frequency of the generated voltage is directly proportional to the machine rotating speed.

2.4.2.2 Equivalent circuit

For a generator connected to a large power system, as depicted in Fig. 28, the generator speed of rotation, N_S, must be governed to remain constant at the value consistent with the power system frequency.[3] With fixed rotation speed, the stator-induced voltage E_a is derived from the DC excitation winding and is adjustable in value by controlling the field current I_F as shown in Fig. 28. Each phase winding of the generator armature may be characterized by the winding resistance R_a and an impedance element known as the "synchronous reactance", X_a. This is not a real physical element but an analytical component arising from sinusoidal flux distribution in the armature winding and magnetic cores of the machine. The current in the stator phase windings will set up magneto motive forces (mmfs) and therefore flux linkages; hence the windings must possess self and mutual inductances across which voltage will be dropped. In the presence of load impedance — in this case, the coupled power system — currents I_a flow in the armature windings. The per-phase voltage V_S at the generator terminals is fixed at the system value and is related to E_a by the phasor equation[15, 18]

$$V_S = E_a - I_a(R_a + jX_a). \tag{58}$$

The use of "j" notation is valid only for quantities varying sinusoidally in time. The term R_a represents per-phase armature resistance in ohms. X_a is the per-phase synchronous reactance in ohms and I_a is the phase current.

Adjustment and control of voltage E_a, shown in Fig. 28, control the reactive power generated. The action of changing the field current and its effect on generator operation is best illustrated in terms of phasor diagrams shown in Fig. 29(a). This demonstrates the condition where an

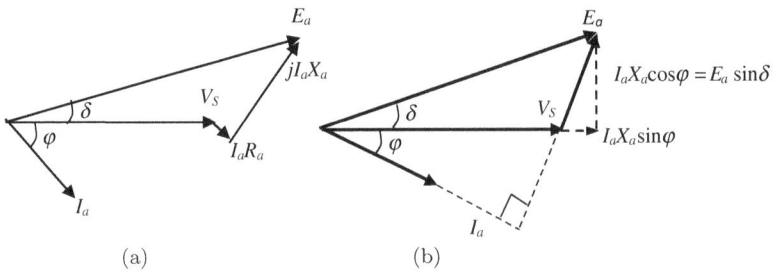

(a) (b)

Fig. 29. Phasor diagram of an overexcited synchronous generator.[5] (a) Delivering a lagging current and (b) delivering a lagging current — neglecting armature resistance.

overexcited cylindrical rotor generator delivers current to the system that lags the terminal voltage V_S in time phase by the angle φ. Usually, the voltage drop $I_a X_a$ is at least ten times the voltage drop $I_a R_a$, so that often the latter is neglected altogether. In the phasor diagram, the voltage component $I_a R_a$ is in time phase with current I_a, while the component $jI_a X_a$ leads I_a in time phase by 90°. The angle δ between the voltage E_a and V_S is usually referred to as "power angle" or "load angle" or "torque angle".

If the armature resistance is neglected, the phasor diagram of Fig. 29(a) can be reinterpreted in more detail in Fig. 28(b). The real power P delivered by the three phase generator is given by

$$P = 3V_S I_a \cos\phi, \text{ and} \tag{59}$$

the reactive power delivered by the generator is given by

$$Q = 3V_S I_a \sin\phi. \tag{60}$$

In Fig. 29(b), it can be seen that

$$E_a \sin\delta = I_a X_a \cos\phi. \tag{61}$$

The term $I_a \cos\phi$ in Eq. (61) can be replaced to give the expression

$$P = \frac{3V_S E_a}{X_a} \sin\delta. \tag{62}$$

Similarly, the reactive power Q of the generator has the magnitude

$$Q = \frac{3V_S}{X_a} [E_a \cos\delta - V_S] \tag{63}$$

When the generator is connected to an infinite bus bar, the terminal voltage V_S is constant. Although the value of the synchronous reactance X_a depends on the state of the generator magnetic saturation, it is almost constant within the normal range of generator operation. From (59) then

$$P \infty I_a \cos\phi, \tag{64}$$

and from (62)

$$P \infty E_a \sin\delta. \tag{65}$$

When the generator is connected to the grid directly (without power electronic converter), variations of the mechanical input power from the

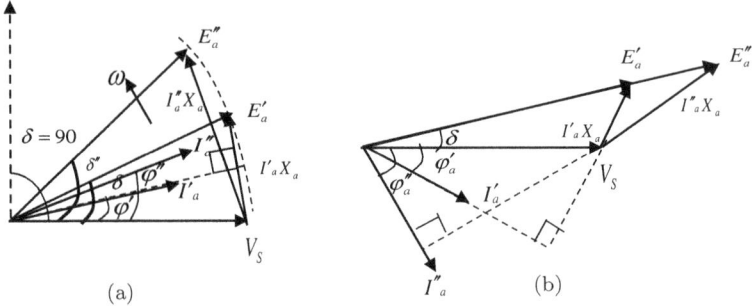

Fig. 30.   (a) Effect of turbine input power change. (b) Overexcited synchronous generator showing the effect of increase of the excitation voltage E_a'.[2,17]

wind turbine due to wind speed changes result in the power provided to the system changing, but V_S must stay the same (infinite bus-bars) and E_a is unchanged (constant field current). The rotor accelerates under the influence of the driving torque and the phasor E_a moves ahead of the phasor V_S until a new steady state is reached when supplied mechanical power is equal to the generated power. As shown in Fig. 30(a), in this case, I_a increases, and so does jX_aI_a. Since X_aI_a must be perpendicular to I_a. So I_a must lead V_S, i.e. we have generated power with a leading power factor angle.

As the wind turbine input power is increased further, vector E_a must move around a circle of constant radius (see Fig. 30(a)), because the field current is not being changed, with $I_a \cos \varphi$ increasing and therefore $X_aI_a \cos \varphi$ increasing. According to Eq. (62), the real power generated increases until it reaches the maximum $P_{\max} = E_aV_S/X_a$ when $\sin \delta = 1$. With further increase of turbine input power, the machine cannot generate equal amount of power to balance the mechanical input power and it accelerates out of synchronism with the supply. It ceases to be a generator synchronized to the system. This is a fault condition and must be rapidly rectified.

An increase of the output power P delivered by the generator is realizable by increasing the variables $E_a, I_a, \cos\varphi$ and $\sin\delta$. But these four variables are interactive — an increase of one will cause a decrease of another. For example, Fig. 30(b) shows the effect of increasing E_a when the torque angle δ is constant. It is seen that the phase $I_a'X_a$ increases to $I_a''X_a$ representing an increase of the armature current delivered from I_a' to I_a''. But this increase is accompanied by an increase of the phase angle from ϕ' to φ'' and a consequent decrease of $\cos \varphi'$ to $\cos \varphi''$. Concurrent increase of the

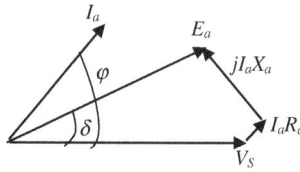

Fig. 31. Under-excited synchronous generator supplying leading current.

delivered current I_a and the phase-angle ϕ is seen from Eq. (50) to result in the increased delivery of lagging reactive power Q. Any change of the state of operation of a generator "on line" must be carefully considered to include the effects on all of the machine electrical variables. In particular, it is necessary to know the effect of any changes on the real power P and the reactive power Q delivered by the machine.

When the magnitude of the induced voltage, E_a, is smaller than that of the terminal voltage V_S, the effect is for the generator to deliver a leading current to the grid (see Fig. 31). The leading volt-amperes delivered to the system can only be absorbed by capacitance within the system, such as that provided by power factor correction capacitors. In effect, the delivery of a leading current from the generator to the power system is equivalent, in terms of reactive power flow, to imposing a lagging (inductive) load. An under-excited synchronous generator is equivalent to a sink of lagging volt-amperes, although it is actually delivering a leading current and is a source of leading volt-amperes. Both the real power P and reactive power Q are invariant with machine speed. Load flow analysis and transient effects are often considered in terms of variations of the load angle δ.[6]

Synchronous machines can also operate as a motor, current flows are possible in four-quadrants. Motor mode is rarely used in wind power generation, hence is not discussed here.

2.5 *Wind power generation systems*

This section gives comprehensive explanations of all the wind-powered generation systems listed in Section 2.1.

2.5.1 *Constant speed, constant frequency (CSCF) wind generators*[19]

For this type of wind generators, despite the variations of wind speed, the rotor speeds are maintained unchanged and determined by the frequency of the utility grid they are connected to, the gear ratio and the generator layout. Many presently installed wind generator systems operate at constant (or near constant) speed, they feature simple and reliable construction of electrical parts, but the mechanical parts are subject to higher stresses. Also, constant speed wind turbines impose certain impacts on the supply grid due to wind speed variations.

Most constant speed wind turbines use squirrel cage induction generators, and many instead use a wound-rotor synchronous generator.

2.5.1.1   Constant speed, constant frequency with squirrel cage induction generator

The simple, robust and highly efficient features of the squirrel cage induction generator have made it popular for use in the constant speed wind turbine systems. The generator operates within a narrow range of rotational slip speeds, which vary according to the generated output power (e.g. slips of about 1–2%). Note the turbine is generally stall controlled. Figure 16 shows a basic configuration where the induction generator is connected directly to the utility grid through two important elements: the capacitor bank and the soft-start switches.[4]

Capacitor bank: This is necessary because an induction generator draws a large excitation current from the grid when connecting, causing high consumption of grid reactive power. This causes low power factor and high losses in the power grid and may result in a large voltage dip. Capacitors are connected in parallel with the individual generator; they supply the reactive power required to establish the rotating magnetic field for voltage induction to take place, and hence improve the grid power factor and reduce losses.

Soft start: Connecting a wind turbine generator directly to the utility grid will cause a large transient inrushing current, hence torque oscillation just as in the case of switching a squirrel cage induction motor onto the supply. To reduce this effect at startup, a three-phase thyristor-controlled AC-AC regulator is typically used as a soft-start mechanism on a constant speed wind turbine. The back-to-back connected thyristor pairs are fired in synchronism with the corresponding supplied phase voltage from $180°$ to $0°$, resulting in a gradual increase of the machine terminal voltage. Adjusting the phase angle should enable the average voltage to slowly increase until the rated flux level according to $V_{rated}/F_{rated} = \phi_{rated}$ is reached. After startup, the thyristors are typically shorted out by mechanical

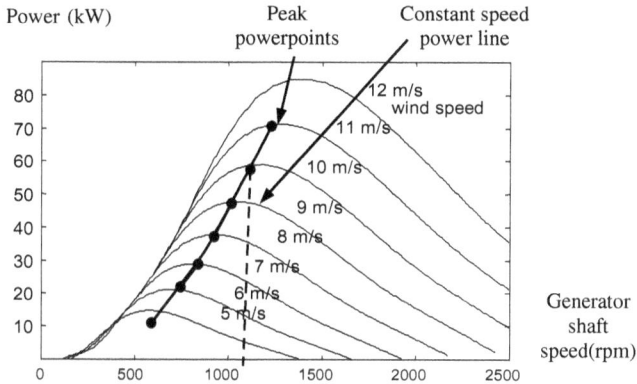

Fig. 32. Constant speed line for a constant speed wind turbine system.[2]

contractors connected in parallel with the thyristor pairs to eliminate power losses.

On the mechanical side, pitch angle control is applied to increase the machine speed to near the synchronous level determined by the pole numbers and grid frequency.

2.5.1.2 Constant speed, constant frequency with wound rotor synchronous generator

This type of machine may be used for a constant speed wind turbine. The stator windings are directly connected to the supply grid, hence the rotational speed is strictly fixed by the frequency of the grid. Figure 32 shows a constant speed line on a set of power–speed curves for a CSCF generator. The main advantage of this machine is that the magnetizing power is provided by the excitation circuit of the rotor, so it does not need any reactive power compensation equipment such as capacitors and batteries.

During transient and sub-transient load steps, the machine may have a problem in staying synchronous with the grid frequency and the only way to help is by controlling the magnetization of the machine. Also, due to strictly fixed speed, the power transient transmitted to the grid becomes more prominent than for its counterpart induction machine.

Strictly fixed speed wind generator systems have a number of shortcomings affecting their widespread application, particularly in large wind farms. These are mechanical stress, reactive power consumption and reduced energy capture.

Mechanical stress: Due to the almost fixed speed of the system, any wind power variations become pulsating torque transmitted along the drive train, gearbox and applied at the generator side. The inflexibilities of the generator speeds result in mechanical stress. To enable the system to withstand the absolute peak loading conditions, safety design mechanisms must be imposed, which add to the cost.

Pitch control may be applied to compensate the effect of wind speed variations on the system by pitching the rotor blades to different angles, and hence changing the power coefficient C_p. However, this mechanism is slow in reacting to wind speed changes and cannot compensate for gusts and high frequency torque pulsations due to blades passing the tower.

Voltage and Power Pulsations: Wind speed variations affect directly the power generated by the fixed speed wind turbines and flowing to the grid, particularly from those with induction generators. These machines have steep slip-torque characteristics, sensitive to speed variations, hence any fluctuations of wind speed will result in power to the grid pulsating. Fixed speed induction generators draw reactive power from the grid, and the reactive power drawn will vary with the real power variations; this causes voltage variations, high losses in the transmission lines, and can be particularly damaging to a weak grid with high source impedance.

Reduction in Energy capture: A wind turbine system should be able to capture as much as possible of the available wind power as long as the wind speeds are varying within the system's rated range. Wind speed changes cause the turbine tip-speed ratio (in Section 1.3) to change, hence moving away from the optimal power coefficient C_p value for the maximum wind power capture. A variable speed wind turbine can regulate the rotor shaft speed corresponding to wind speed changes, but a constant speed system cannot. Consequently, the energy captured by constant speed wind turbine systems are lower compared to the variable speed systems of similar power ratings.

2.5.2   Variable speed singly-fed wind generators

To overcome the problems related to constant speed wind generators, most modern wind farms install variable speed, variable frequency wind power generators. There are several different types of variable speed systems as shown in Figs. 18–19, differing in the types of generator and power electronic converter used.

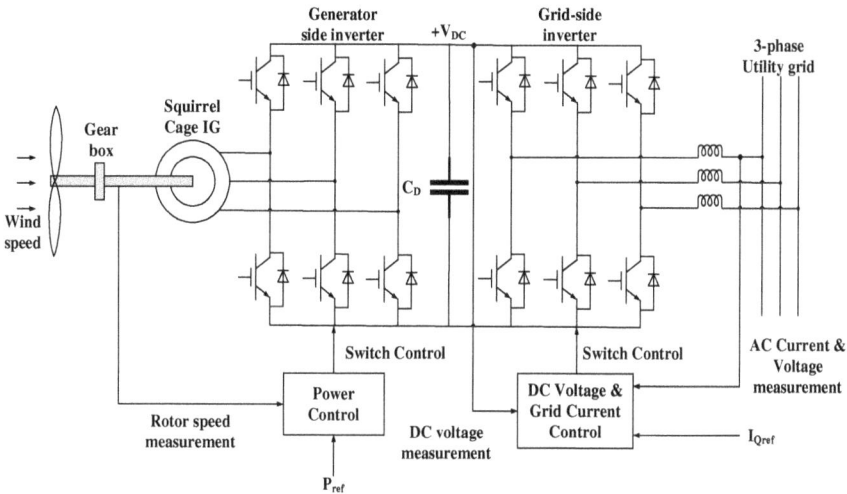

Fig. 33. Variable Speed induction generator controlled by stator-side AC-DC-AC converter.

2.5.2.1 Variable-speed constant frequency (VSCF) induction generators

An AC-DC-AC converter can be connected between generator stator windings and power network phase terminals as shown in Fig. 33 The frequency and voltage of generator-side converter can be flexibly controlled to magnetize the machine and then enable power generation from the wind.

Stator Voltage Control: Field excitation to the generator is performed just like the case when it is directly connected to the grid. In this case, the same soft start as that applied to the constant speed wind turbine systems is used. Then generator speed or power control is implemented according to the measured wind speed and a dedicated control scheme. This will adjust the stator-side converter voltage and frequency, and hence regulate the electromagnetic torque, subsequently changing the generator rotor speed. Vector control or direct torque control scheme[19-21] may be applied.

Grid-side power Control: The grid-side converter performs the function of controlling the active and reactive power to the grid. By keeping the DC-bus voltage constant, it defines the reference current value corresponding to the active power feeding to the grid. For unity power factor of the current flow to the grid, the reference value for reactive current may be set to zero, hence no reactive power is injected to/from the grid. This converter allows bidirectional power flow. However, the converter rating needs to be

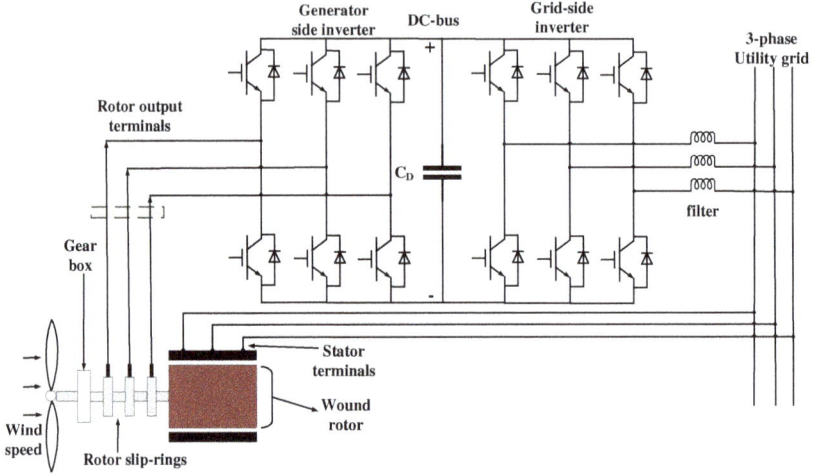

Fig. 34. Converters for grid connected doubly fed wound rotor induction generator.

approximately 125% of the machine rating as the reactive power required by the machine must be supplied by the converter.

2.5.3 *Variable speed doubly fed induction generators*

The doubly fed induction generator (DFIG) system can now be regarded as the standard solution of today. Over 85% of installed wind turbines utilize DFIGs[7, 10] and the largest capacity in the industry for a commercial wind turbine product with DFIG has increased towards 5MW. A typical DFIG configuration is as shown in Fig. 34.

The stator winding is directly connected to the grid, in some cases through a transformer, while the rotor windings are connected to the grid through an AC-DC-AC converter. This has an intermediate DC bus and a capacitor as energy storage like a conventional static Kramer system.[2, 18] The machine rotor-side converter may be an uncontrolled diode bridge rectifier, but now mostly uses a PWM controlled AC-DC converter; the grid side is a three-phase DC-AC inverter generally connected to the grid via a transformer for voltage level adaption. The converter is designed for slip power conversion up to ±30% of the stator power rating. With the rotor-side converter, it is possible to control the torque or speed of the DFIG and also the power factor at the stator terminals. The main objective for the grid-side converter is to keep the DC-bus voltage constant regardless of the magnitude and direction of the rotor power. The grid-side converter works at the grid frequency (leading and lagging in order to absorb a controllable level of reactive power). The rotor-side converter works at different frequencies, depending on the machine rotor speed according to wind speed.

2.5.3.1 Principles of DFIG operation

With stator windings connected directly to the AC power network, the frequency of the stator three-phase voltage is constant and equal to the network frequency f_s. This requires that the air-gap magnetic field passing through the generator stator windings is rotating at the constant speed determined by the grid frequency, even though the generator rotor speed n_r varies, due to fluctuations of the mechanical power provided by the wind turbine. To achieve this, the frequency f_r of the AC current fed to/drawn from the rotor windings of the DFIG generator must be continuously adjusted to counteract any variations in the rotor speed. Thus for the constant rotating field frequency f_s, and the variable rotor rotation speed, n_r(rpm), f_r determined by the rotor-side converter frequency can be calculated as

$$f_r = f_s - \frac{n_r \times p}{120},$$

where p is the number of poles. According to rotor speed in f_r (rad/sec) in comparison to synchronous speed f_S, a DFIG has three operation modes. Using angular speed $\omega = 2\pi f$(rad/sec), these are:

$\omega_{\mathbf{r}} > \omega_{\mathbf{s}}$, **super-synchronous mode**, slip $s < 0$, $f_r < 0$. The rotor slip power is fed to the grid, the generator rotor-side converter acts as a rectifier, and the grid-side converter as an inverter. Stator active power is fed to the grid.

$\omega_{\mathbf{r}} < \omega_{\mathbf{s}}$, **sub-synchronous mode**, slip $s > 0$, $f_r > 0$. In this mode, rotor slip power is supplied by the grid through the generator side AC-DC converter which acts as an inverter. The grid-side converter acts as a rectifier, and active power is fed from the grid to the rotor. With sufficient wind turbine input power, the stator can still generate power and feed it to the grid.

For $\omega_{\mathbf{r}} = \omega_{\mathbf{s}}$, **synchronous mode**, slip $s = 0$, so $f_r = 0$. This makes the machine operate like a conventional synchronous machine; the grid supplies DC current to the rotor for field excitation, and power generated is fed to the grid through stator.

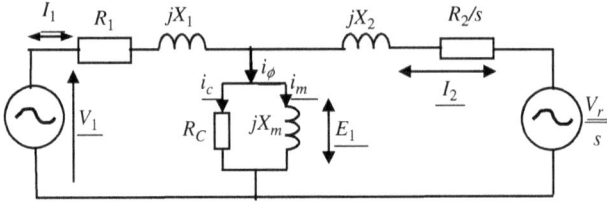

Fig. 35.   Per-phase equivalent circuit for a DFIG with a voltage source inserted in the rotor circuit.

2.5.3.2   Per-phase equivalent circuit

The per-phase equivalent circuit of a DFIG is shown in Fig. 35.[22] It differs from that of a conventional induction machine shown in Fig. 21(a) since a voltage source V_r is added into the rotor side. Ignoring magnetizing core loss R_C, and viewing from both stator and rotor ends, respectively, the stator and rotor-side voltages can be expressed as

The stator-side voltage equation can be expressed as

$$\underline{V_1} - \underline{E_1} = \underline{I_1}(R_1 + j(X_1)), \tag{66}$$

and that for the rotor-side is

$$\underline{V_r} - s\underline{E_1} = \underline{I_2}(R_2 + jX_2'), \tag{67}$$

where $X_2' = s\omega_2 L_2 = sX_2$. Dividing by s on both sides of Eq. (67) gives

$$\frac{\underline{V_r}}{s} - \underline{E_1} = \underline{I_2}(R_2/s + j\omega_2 L_2) = \underline{I_2}(R_2/s + jX_2). \tag{68}$$

Terms $\frac{R_2}{s}$ and $\frac{V_r}{s}$ can be clearly separated into slip-dependent and loss-dependent terms as

$$\frac{R_2}{s} = R_2 + \frac{R_2(1-s)}{s}, \tag{69}$$

and

$$\frac{V_r}{s} = V_r + \frac{V_r(1-s)}{s}. \tag{70}$$

Subsequently, the per-phase equivalent circuit shown in Fig. 35 can be re-drawn to show the loss terms and slip related terms in the rotor circuit as that shown in Fig. 36.[23]

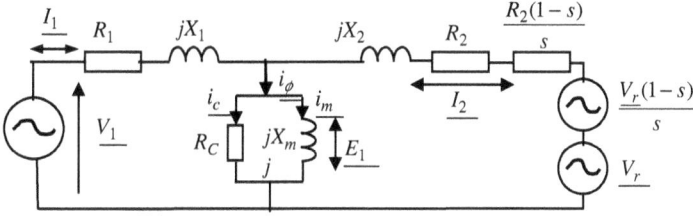

Fig. 36.   Per-phase equivalent circuit for a DFIG with loss and slip-related terms shown in the rotor circuit.

2.5.3.3   Mechanical input power and torque

The mechanical power, P_{mech}, in/out of an induction machine is the power associated with the slip-dependent terms $R_2(1-s)/s$ and $Vr(1-s)/s$. This can be verified from the power balance relation:

$$P_{mech} = P_s + P_r - P_{loss,s} - P_{loss,r}. \tag{71}$$

In motoring operation, P_s and P_r are powers supplied to the machine from the electric power source through the stator and rotor windings, respectively, and $P_{loss,s}$ and $P_{loss,r}$ are, respectively, the stator and rotor winding losses, so P_{mech} is the mechanical load power which is less than the supplied input electric power. In the case of generators, all 4 power terms in Eq. (71) are from the turbine mechanical driving power, and P_s and P_r are negative, by motor convention, so the magnitude of P_{mech} is greater than the generated electric power fed to the grid.

Expressing the right-hand-side terms of the power balance Eq. (71) in terms of the circuit parameters using the slip-dependent terms, P_{mech} is written as

$$P_{mech} = 3\left|I_2^2\right| \left(\frac{R_2(1-s)}{s}\right) - 3\mathrm{Re}\left\{\underline{V}_r\left(\frac{1-s}{s}\right)\underline{I}_2^*\right\}$$

$$= 3\left(\frac{1-s}{s}\right)\left\{\left|I_2^2\right| R_2 - \mathrm{Re}\left\{\underline{V}_r\underline{I}_2^*\right\}\right\}, \tag{72}$$

where I_2^* is the complex conjugate of the rotor current phasor $\underline{I}_2$ and assuming the simplified circuit, the current magnitude can be written as

$$\left|\underline{I}_2\right| = \frac{\left|V_1 - V_r/s\right|}{\sqrt{(R_1 + \frac{R_2}{s})^2 + (X_1 + X_2)^2}}. \tag{73}$$

With the mechanical power input expressed as in Eq. (72), we can derive the torque as

$$T_m = P_{mech} \frac{p}{\omega_r} = 3 \frac{p}{\omega_r} \left(\frac{1-s}{s} \right) \left\{ |I_2^2| R_2 - Re \left\{ \underline{V}_r \underline{I}_2^* \right\} \right\}, \qquad (74)$$

where ω_r is the rotor speed in rad/sec, p is the number of pole pairs. Since

$$\frac{1-s}{s} = \frac{\omega_r}{\omega_s - \omega_r} = \frac{\omega_r}{\omega_m}, \qquad (75)$$

where ω_m is the slip speed, $\omega_s - \omega_r$. So the torque equation can be further simplified as

$$T_m = P_{mech} \frac{p}{\omega_r} = 3 \frac{p}{\omega_r} \left(\frac{\omega_r}{\omega_m} \right) \left\{ |I_2^2| R_2 - Re \left\{ \underline{V}_r \underline{I}_2^* \right\} \right\}$$

$$= 3 \frac{p}{\omega_m} \left\{ |I_2^2| R_2 - Re \left\{ \underline{V}_r \underline{I}_2^* \right\} \right\}. \qquad (76)$$

The real part of the 2nd term on the RHS represents the real power flow through the rotor circuit and can be rewritten in scalar form as

$$T_m = \frac{3p}{\omega_m} \left\{ I_2^2 R_2 - V_r I_2 \cos(\phi_v - \phi_i) \right\}. \qquad (77)$$

Note φ_v is the phase angle between voltage phasor $\underline{V}_1$ and $\underline{V}_r$, and φ_i is the phase angle between $\underline{V}_r$ and $\underline{I}_r$.

The real power supplied to the grid through the stator circuit is

$$P_s = 3 Re \left\{ \underline{V}_1 \underline{I}_1^* \right\}. \qquad (78)$$

According to Eq. (66) and $\underline{E}_1$, expression, the stator voltage phasor

$$\underline{V}_1 = \underline{I}_1 (R_1 + jX_1) + (\underline{I}_1 + \underline{I}_2) jX_m. \qquad (79)$$

Substituting $\underline{V}_1$ into (78), the stator power can be derived as

$$P_s = 3 R_1 I_1^2 + 3 Re \left\{ jX_m \underline{I}_2 \underline{I}_1^* \right\}. \qquad (80)$$

Similarly, the rotor power can be expressed as

$$P_r = 3 Re \left\{ \underline{V}_r \underline{I}_2^* \right\}. \qquad (81)$$

According to Eq. (67), and expressing $s \underline{E}_1$ in terms of circuit parameters, rotor voltage phasor is written as

$$\underline{V}_r = \underline{I}_2 (R_2 + jsX_2) + (\underline{I}_1 + \underline{I}_2) jsX_m. \qquad (82)$$

Substituting $\underline{V}_r$ into (81), the rotor power is

$$P_r = 3R_2I_2^2 + 3\text{Re}\left\{jsX_m\underline{I}_1\underline{I}_2^*\right\}. \tag{83}$$

2.5.3.4 DFIG power balance

The power transfer relationship is similar to that described previously. According to Eq (71), the power across the air-gap to the stator is called air-gap power $P_{ag} = P_s - P_{loss,s}$ (note $-$ve P_s for generator.). The turbine mechanical input power P_{mech} can be written in terms of P_{ag} as

$$P_{mech} \approx P_{ag}(1 - s) = P_{ag} - sP_{ag}. \tag{84}$$

(Note for generator mode, P_{ag} is — ve, and slip s can be –ve.) So P_{mech} is the sum of air-gap power and the rotor power P_r, which is written as

$$P_r = P_{mech} - P_{ag} = P_{ag}(1 - s) - P_{ag} = -sP_{ag}. \tag{85}$$

The generator rotor power P_r may be drawn from or supplied to the utility grid depending on the sign of slip s. Negative sign of the per unit slip in Eq. (83) P_r is negative (due to negative P_{ag}) indicating the rotor power flowing to the grid.

Neglecting the stator copper loss and core loss, the stator power is approximately equal to the air-gap power $P_s \cong P_{ag}$. The rotor power, Eq. (85), can be written as $P_r \cong sP_s$. The electrical output power P_{net} of IG is in turn equal to $P_s + P_r$, giving as

$$P_{net} = (1 - s)P_{ag} = P_s + P_r. \tag{86}$$

Alternatively, P_{net} can be represented in terms of the resultant input power of the system and the total electrical losses P_{eloss} as:

$$P_{net} = |(1 - s)P_{ag}| - P_{eloss} = |P_s + P_r| - P_{eloss}, \tag{87}$$

where $P_{eloss} = $ stator copper loss + core loss + rotor copper loss + converter loss. Figure 37 shows a power flow diagram of a DFIG.

To summarize, using a doubly fed induction generator instead of an asynchronous generator in wind turbines offers the following advantages:

- operation at variable rotor speed while the amplitude and frequency of the generated voltages remain constant;
- optimization of the amount of power generated for a given available wind up to the nominal output power of the wind turbine generator;
- virtual elimination of sudden variations in the rotor torque and generator output power;

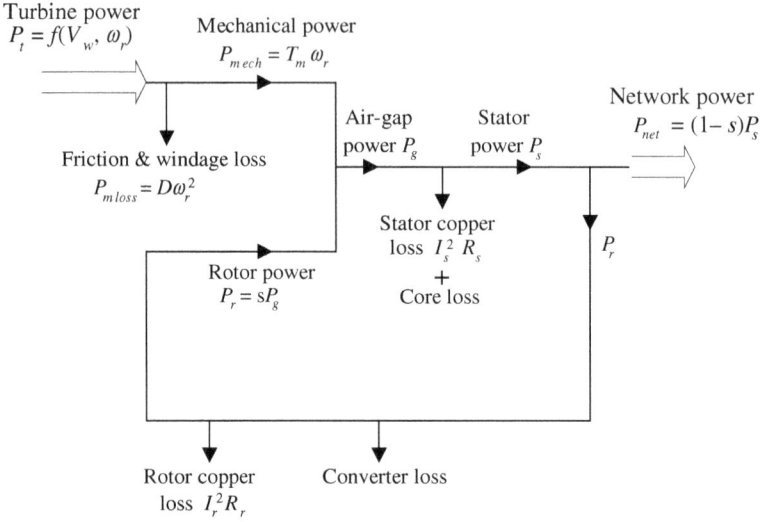

Fig. 37.   Power flow diagram for a doubly fed induction generator.[2,7]

- generation of electrical power at lower wind speeds;
- control of the power factor (e.g. in order to maintain the power factor at unity).

    On the other hand, the doubly fed induction generator requires complex power conversion circuitry which the asynchronous generator does not need. Also, the slip rings on the wound-rotor induction machine, used to implement the doubly fed mechanism, require periodic maintenance while no such rings are required on the rotor of the squirrel-cage induction machine used to implement the asynchronous generator.

References

1. Horizontal-Axis Wind Turbine (HAWT) Working Principle | Single Blade, Two Blade, Three-Blade Wind Turbine, Electrical Academia.com.
2. W. Shepherd and L. Zhang (2017). *Electricity generation by Wind Power*. Singapore: World Scientific.
3. W. Shepherd and D. Shepherd (2003). *Energy Studies*, 2nd ed., Chapter 10. London, England: Imperial College Press.
4. Siegfried Heier (2006). *Wind Energy Conversion Systems*, 2nd ed. Chichester: Wiley.
5. Volker Quaschning, James (2016). *Understanding Renewable Energy Systems*. Routledge.

6. Li Zhang (2017). *Electricity Generation by Renewable Energy Systems*, MSc course notes, University of Leeds, UK. http://www.wind-power-program. com/popups/powercurve.htm.

7. https://en.wikipedia.org/wiki/Technical_University_of_Denmark.

8. https://en.wikipedia.org/wiki/Neodymium.

9. Dimitris Bourlis (2011). *A complete Control Scheme for Variable Speed Stall Regulated Wind Turbines*. Online publication, July 5, 2011. DOI:10.5772/17569.

10. Z. Lubosny (2003). *Wind Turbine Operation in Electric Power Systems*, Chapter 5. Heidelberg, Germany: Springer-Verlag.

11. T. Man, J.P. Sullivan, and O. Wasynczuk (1981). Dynamic Behaviors of a Class of Wind Turbine generators During Random Wind Fluctuations. *IEEE Transactions on Power Apparatus and Systems*, **June**, PAS100(6).

12. S.M. Muyeen, Md. Hasan Ali, R. Takahashi, T. Murata, J. Tamura, Y. Tomaki, A. Sakahara, and E. Sasano (2007). Comparative study on transient stability analysis of wind turbine generator system using different drive train models. *IET Renew. Power Gener.*, **1**(2), 131–141.

13. Patil Ashwini and Thosar Archana (2016). Mathematical Modeling of Wind Energy System Using Two Mass Model Including Generator Losses. *International Journal of Emerging Trends in Electrical and Electronics*, **12**(2), 18–23.

14. Niels I. Meyer (1995). Danish wind power development. *Energy for Sustainable Development*, **II**(1), 18–25.

15. Electric Machines (ELEC3565) (2019). BSc Electronic and Electrical Engineering Degree Course Notes, University of Leeds, UK.

16. https://www.polytechnichub.com/types-induction-motor/.

17. P.L. Alger (1970). *Induction Machine*, 2nd ed., New York: Groden and Breach Science Publishers.

18. G. McPhaerson (1990). *An Introduction of Electrical Machines and Transformers*, 2nd ed. New York: John Wiley & Sons.

19. H. Sloorweg and E. de Vries (2003). Inside wind turbines (fixed and variable speed). *Renewable Energy World*, **6**(1), 31–40.

20. D.W. Novotny and T.A. Lipo (1996). *Vector Control and Dynamics of AC Drives*. New York: Oxford Science Publication, Oxford University Press.

21. P. Vas (1990). *Vector Control of AC Machine*. New York: Oxford Science Publishing.

22. J.M.D. Murphy and F.G. Turbull (1988). *Power Electronic Control of AC Motors*. Oxford, UK: Pergamon Press.

23. Wenping Cao, Ying Xie, and Zheng Tan. Wind Turbine Generator Technologies. Advances in Wind Power, Rupp Carriveau, IntechOpen, DOI: 10.5772/51780. Available from: https://www.intechopen.com/books/advances-in-wind-power/wind-turbine-generator-technologies, 50.

24. J. Microw. Wound rotor doubly fed induction machine with radial rotary transformer, Optoelectron. Electromagn. Appl. vol.12 no.2 São Caetano do Sul Dec. 2013. https://doi.org/10.1590/S2179-10742013000200013.

Chapter 3

Costing and Future Development of Wind Turbine Technology

Li Zhang[*,‡] and Qiou Yang Zhou[†,§]

[*]*The University of Leeds, Leeds, UK*
[†]*South West Jiao Tong University of China, Sichuan, China*
[‡]*l.zhang@leeds.ac.uk*
[§]*ukqiuyang@163.com*

Abstract

The first part of this chapter provides an extensive review of commercial aspects and costs of establishing and running a wind power installation (or "wind farm"). The final section then discusses the future development of wind turbine technology, which constantly aims to make turbine blades longer, lighter, stronger and more cost effective. Composite materials are a main enabler here, and are discussed in some detail.

1 Major Cost Items of Wind-Powered Electricity Generation

Like any energy supply system, wind-energy systems incur both installation costs and operating costs. The factors that influence the costs of wind-powered electricity generation include:[1]

- the cost of the turbines and generators,
- the cost of the turbine site, construction, and grid connection,
- the operation and maintenance costs during the system lifetime,
- the turbine lifetime and depreciation rate,
- the business costs associated with the financing of the building and operation,
- the wind regime at the site,

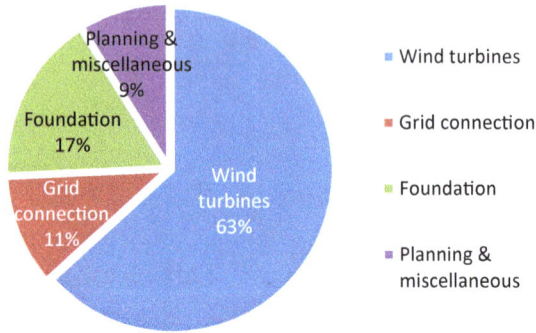

Fig. 1.   Typical wind farm construction expenditure.[2]

- the energy capture efficiency of the turbines, and
- the availability for sale of the generated electricity.

Wind-energy project capital costs, reported by the International Energy Agency, show substantial variation among countries. This is due to such factors as the market, the physical structures, site characteristics and planning regulations. The installation cost can be expressed as a function of the rated electrical capacity,[1] and economies of scale are expected.

Figure 1 shows the percentage contribution of each cost element to the installation expenditure of a typical wind farm. Clearly the turbine and generator, treated as one item, usually take the largest share of the total, being between 63–75%.

2   Levelized Cost of Electricity (LCOE)

To assess and compare the costs of wind power and other electricity generating techniques, an important and commonly used metric, the Levelized Cost of Electricity (LCOE),[3] is used.

The LCOE allows evaluation of the per megawatt hour price of a generating technique over its lifetime.[3] It has been applied to electrical energy produced by a variety of sources; it can be applied to any renewable-sourced form of generation including wind. For a project developer, the LCOE gives an estimate of the wholesale power price required to cover all the expenditures at the commissioning stage and obtain a desired profit return. The simplified LCOE calculation formula[3] is given as follows:

$$LCOE = \frac{(CapEx \times FCR) + OpEx}{(AEP_{net}/1000)}$$

where:

$LCOE$ = levelized cost of energy ($/megawatt-hour [MWh])
FCR = fixed charge rate (%)
$CapEx$ = capital expenditures ($/kilowatt [kW])
AEP_{net} = net average annual energy production (MWh/megawatt [MW]/ year [yr])
$OpEx$ = operational expenditures ($/kW/yr).

The first three items in the above equation, namely CapEx, OpEx and AEP, can capture system level impacts from design changes (e.g. larger rotors or taller wind turbine towers). The fourth input, the fixed charge rate (FCR), represents the amount of revenue required to pay for the carrying charges as applied to the CapEx on that investment during the expected project economic life on an annual basis.

When applying LCOE to any power plant, the whole expenditure calculation will include its installation, operation and maintenance, its expected life span, the project scale, risk, return and capacities. For a wind power system, the LCOE is determined by the capital costs, the financing costs, the cost of operation and maintenance and the predicted power production. Onshore and offshore wind farms obviously have inherently different cost breakdowns.[3]

Figure 2 shows the annual global weighted average levelized cost of onshore wind-generated electricity for years 2014–2017. It is clear the LCOE

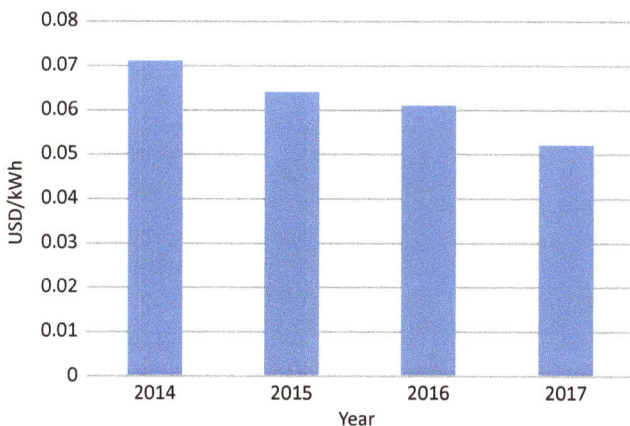

Fig. 2.   The global weighted average levelized cost of electricity from onshore wind farms.[4]

decreases continuously, from $0.07/kWh in 2014 to $0.05/kWh in 2017. The
records appear to show over time that a doubling of the cumulative onshore
installed capacity coincides with a 15% decline in the LCOE of onshore
wind projects. Causality is not evident; it is known that a decreasing cost
for operation and maintenance over time was involved in this downward
trend, but there was also the influence of reduced capital cost, due to the
increasingly mature technology and a more comfortable financial market
for wind power.

For offshore wind farms, although the total installed costs of offshore
projects increased around 8% between 2010 and 2016, the offshore projects'
global weighted average LCOE decreased from $0.17/kWh to $0.14/kWh.
The application of improved technology, which allows higher capacity factors that offset the growth of installed cost, achieves this result.[4]

3   Wind Turbine/Generator Cost

The capital cost of a wind turbine/generator unit is by far the most expensive item in setting up a wind farm and is measured by $/kW. A wind
generator consists of turbine rotor blades and rotor, a gearbox, a tower,
generator, power converter and possibly a transformer. Figure 3 shows the
cost share of each element of a wind generation system. It can be seen that
more than a half of the total expenditure for a wind generator is contained
in the rotor blades, tower construction and the gearbox.

The average wind turbine prices over a recent four-year period are
shown in Fig. 4, revealing a steady downward trend. The price in 2014

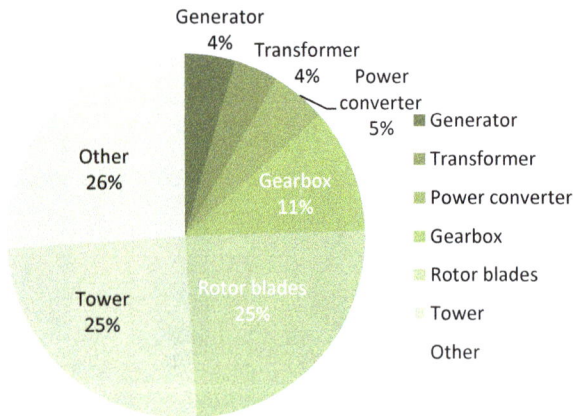

Fig. 3.   Wind turbine cost distribution.[5]

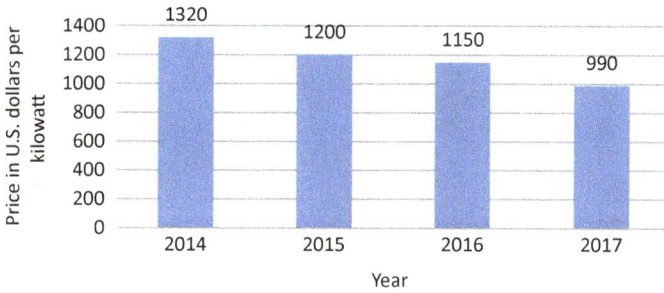

Fig. 4.   The per kilowatt price of wind turbines[6] (rotors larger than 95 meters).

was about \$1320 per kW, and in 2017 it dropped by about \$330/kW to \$990/kW. This was a reduction of about 25% in four years.

In the statistics researched by Bloomberg New Energy Finance (BNEF), wind turbines were divided into two categories, with rotor diameters above or below 95 meters. As shown in Fig. 4, for the former, the prices decreased from \$1100/kW in 2014 to \$950/kW in 2018, being around a 53% reduction compared with that of the last decade, while the price for smaller turbines decreased around 41%. This value is in line with the decline observed in the average selling price for Vestas wind turbines over the period, at 48%, and close to values observed in the United States, for the vast majority of contracts.

This is due primarily to technical improvement. The gradual innovation in both design and operation of the turbine includes growth in the turbine blade lengths, and consequently the hub-height, swept area and average power rating. With these improvements working together, the cost of wind power generation in \$/kW is decreasing steadily.

The other important factor spurring price reduction is market competition, as the number of wind turbine manufacturers increases. The most prominent manufacturers at present are General Electric (USA), Siemens (Germany) and Vestas (Denmark). These make a wide range of wind turbine products, typically more than 20 different models each. This is beneficial to customers who can choose the most suitable design and scale of turbine with the lowest LCOE according to the planned wind farm location. Also, since over a half of the components of a wind turbine are the same, customers can use the same structural components having the lowest cost. This clearly decreases the development expenditure and improves supply chain efficiency.

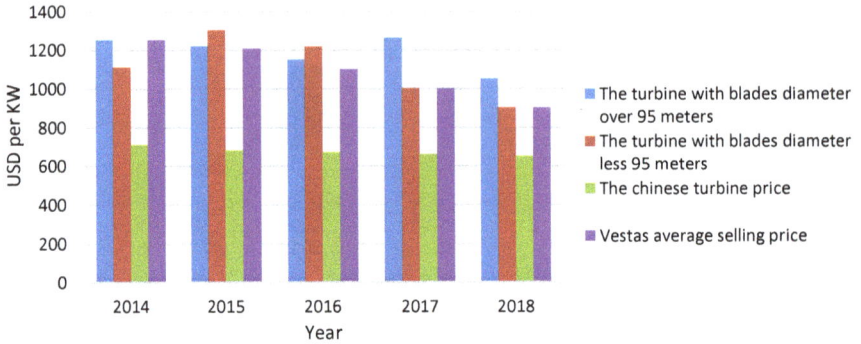

Fig. 5. Wind turbine price indices and price trend.[4]

This downward price trend will continue into the future, due mainly to the increased competitiveness of Chinese manufacturers. Green bars in Fig. 5 show that Chinese-made turbines are significantly cheaper than those made by Vestas (shown in orange bars). Penetration of the established market is currently relatively modest, but Chinese manufacturers could make more contribution to reducing wind turbine costs in the future.

4 Costs of Planning, Construction and Grid Connection

Building a wind farm site involves the construction of foundations for the turbine towers, the transformers and the switching stations. Any access and service roads, as well as buildings to house the instrumentation, control and telecommunication equipment, must also be constructed. Obviously, the associated costs can differ substantially for onshore and offshore installations.

4.1 *Onshore wind farm*

Apart from the costs of the turbine/generator which account for 64% of the total, the rest of the total cost for an onshore wind farm is the site construction, as shown in Fig. 6.

It is worth noting that the cost variations of renewable energy projects could reflect variations in a range of individual costs, for instance, site location characteristics of specific projects. This includes the amount of infrastructure required to access the site, the distance between the site and manufacturing center, the distance between the site and grid interconnected

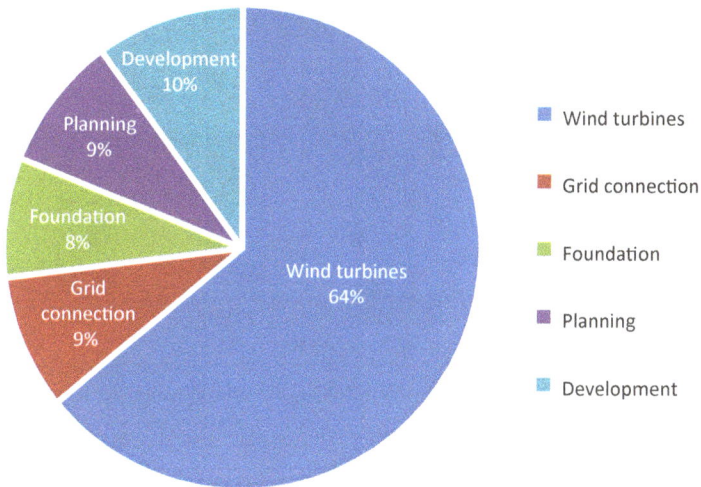

Fig. 6.   Onshore wind farm construction expenditure.[7]

point, and the cost of labor. The costs can be categorized as

- *Foundation construction*, which takes 8% of the total. This is determined by the preparation and construction of the project and the cost of materials and building work for the towers.
- *Grid connection costs*, which are related to the transformers, substations and connections to the local distribution or transmission network. This cost is about 9%.
- *The planning and project costs*, which come from development fees, the related licenses, financial closing costs, feasibility study, legal expenditure, owners' insurance, debt service reserve and construction management.
- *The land cost*, which shares the smallest part of the total cost. In order to reduce the high administrative spending related to the ownership of land, wind farm builders usually lease land through long-term contracts. However, sometimes purchasing land directly is the best choice.

4.1.1   *Total cost of an onshore wind farm*

This covers all costs for building an onshore wind farm, including costs of turbines and generators. The global average installation cost of onshore

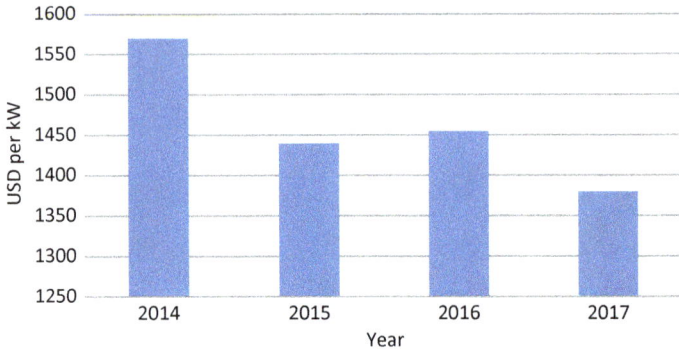

Fig. 7.   Total costs of onshore wind farms — global weighted average.[8]

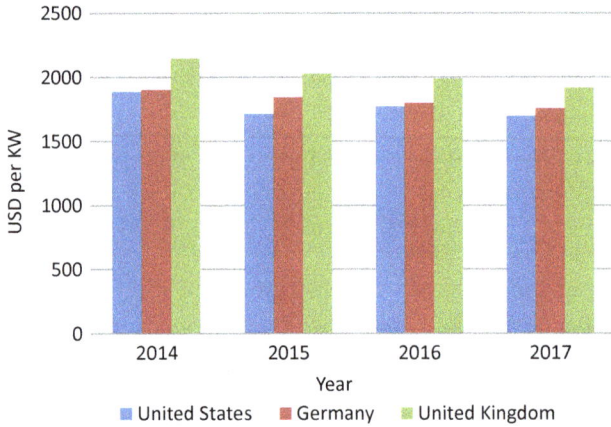

Fig. 8.   Onshore wind farm installation weighted average total costs.[4]

wind farms decreased around 70% over the period between 1983 and 2017, from 4880 to around 1380 USD per kW. Figure 7 presents the cost changes from 2014 to 2017.

Naturally, the cost reductions differ between countries and regions. Figure 8 shows average total installed costs of onshore wind farms for three countries: United States, United Kingdom and Germany, over four years from 2014 to 2017. Clearly the cost in UK is the highest followed by Germany. The United States has the most competitive cost among the three countries, accounting for 1775 USD per kW in 2016. To some extent, this reflects the differences between the relative economic status of these countries.

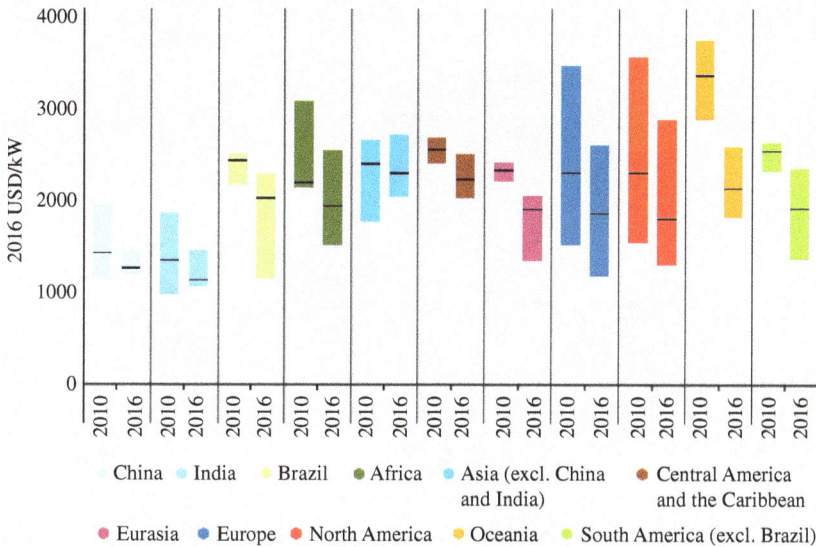

Fig. 9.   Total installation cost ranges and weighted averages for onshore wind farms by country or region, between 2010 and 2016.[4]

Figure 9 shows the total installed cost ranges and weighted averages for onshore wind farms over a number of countries or regions between 2010 and 2016. As can be seen, China and India are the countries having the lowest installed costs with their respective weighted average totals estimated to be USD 1,245/kW and USD 1,121/kW in 2016, which translates into a decline of 11% and 16%, respectively, from 2010. Weighted average installed costs have declined in Brazil from USD 2,390/kW in 2010 to 1,994/kW in 2016. In terms of regions, Asia (excluding China and India), Oceania, Central America, the Caribbean and South America (excluding Brazil) are the most expensive regions, with weighted averages of between USD 1,884 and USD 2,256/kW in 2016. North America has competitive onshore wind farm installation costs, with a weighted average of USD 1.775/kW in 2016. Between 2010 and 2016, costs fell by 36% in Oceania, 22% in North America, 19% in Europe and between 13% and 19% in other regions.

4.2   Offshore wind farms

In comparison with onshore wind farms, building offshore wind farms requires much greater lead times, and the cost is higher and heavily location-dependent. Although the cost of the wind turbine still accounts for the

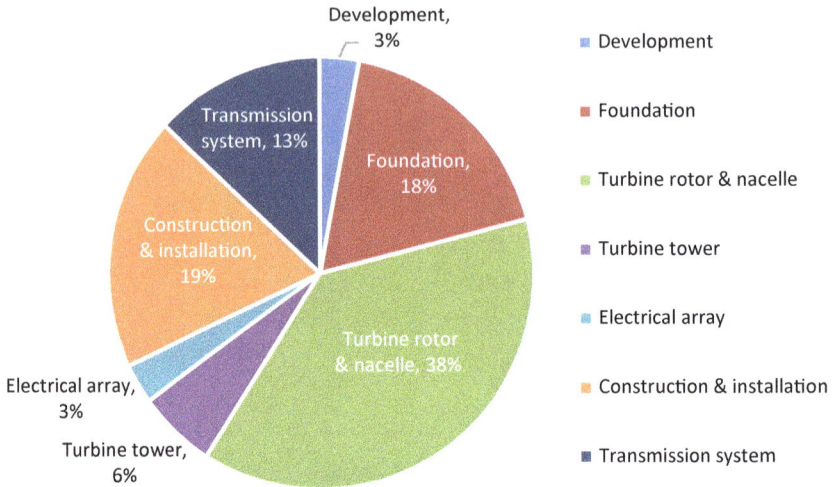

Fig. 10.   Offshore wind farm construction expenditure.[3]

largest proportion of the total, its percentage is lower than that of onshore wind farms, reaching between 30% and 50% of the total. This is because the planning and building of an offshore wind farm, and its constructional components, are much more complex and expensive than for an onshore installation. The other part of the total cost for an offshore wind farm is the site construction, as shown in Fig. 10.

4.2.1   *Breakdown of cost*

In detail, the construction expenditures are compared with onshore wind farms as follows:

- ***Foundation cost***; this varies between 8% and 18%. As is to be expected, spending on foundations shares a significant part of total costs, due to the cost of offshore operations and structural design needed for the harsh marine environment. The selection of the foundation, and hence the cost, depends on factors such as the depth of water, the seabed status, the mass of the rotor and nacelle, the loading and the rotor speed. On the other hand, it is impacted by the familiarity of the company with various options and the company's related expertise, and the supply chain capability in both production and installation. Until now, the utility scale wind farms only used fixed foundations, whereas floating offshore wind farms are under development.[9]

- **The cost of installation and construction**; this occupies around 19% of the total. Purpose-built jack-up vessels are used to carry out the foundation as well as the turbine installations. Floating heavy lift vessels with a dynamic positioning system can also be used to carry out the foundation installation. The turbine installation procedure is sensitive to strong winds, resulting in serious delays and overrun cost. When the wind speed is higher than 13 m/s, the installation has to be halted. This, in turn, results in cost increases.
- **Grid connection** represents the building of the power transmission system for the generated electricity, which increases from 9% to 13%. The distance from the turbine to the port significantly affects the spending on the transmission system. Underwater cable transmission is much more difficult than transmission of electrical power on land due to the complex marine environment and the effect of underwater pressure on the cable. Underwater cables are also likely to be longer.
- **The development and planning costs** decrease in the range between 3–9%. However, this is because of the dramatic increase in other expenditures, resulting in a relatively small proportion of the total.

4.2.2 *Total cost of offshore wind farm*

Overall, the global total installed cost of offshore wind projects has decreased continuously, as shown in Fig. 11. In 2014, a typical offshore wind farm cost around 4760 USD per kW, and by 2017 it was down to 4430 USD per kW. This price is clearly much higher than for the onshore wind farm. Several reasons for this have already been mentioned, and an offshore wind farm must also be able to withstand harsh marine environments. Not surprisingly, offshore operating and maintenance costs are higher. Costs in individual cases may be affected by consideration of a variety of factors, including geographic location and water depth.

5   Operation and Maintenance (O&M) Costs[4]

In calculating the operation and maintenance costs of wind farms, there is an uncertainty. Full-service contracts have increased from $14/kW/year in 2008 to $30/kW/year in 2017. Full-service renewal contracts grew from $22 to $44/kW/year. The O&M costs in the United States were between $16 and $37 kW in 2016. The weighted average of O&M cost was $27/kW/year. This cost increase occurs for several reasons, such as the limitation of turbines' service lifetime and the equipment of wind farms beginning to age.

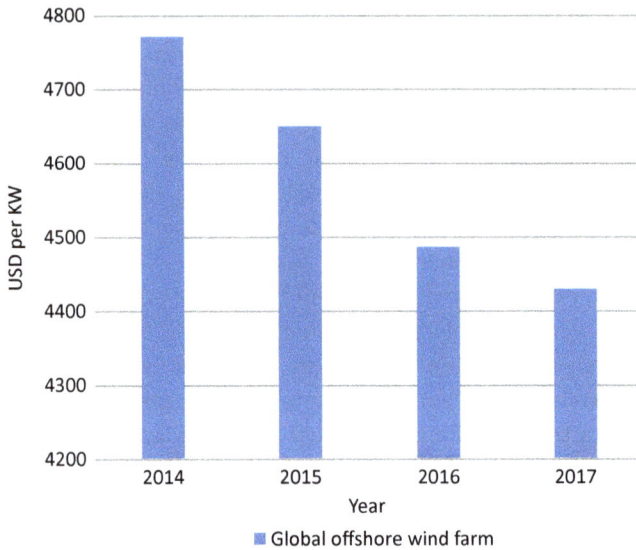

Fig. 11. The evolution of total installed costs for offshore wind farms.[10]

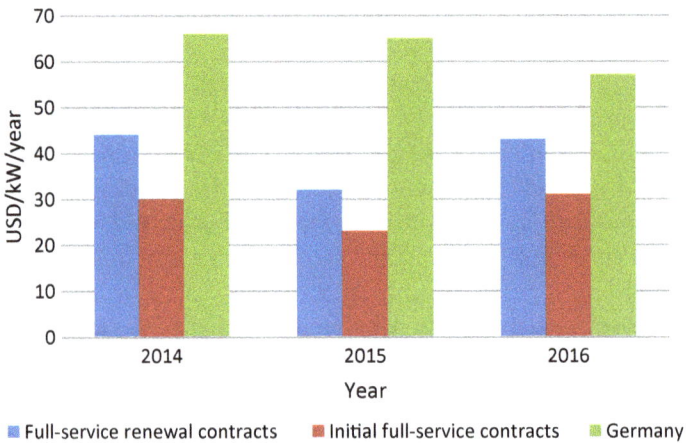

Fig. 12. Full-service (initial and renewal) and Germany O&M cost.

Although the deployment of wind projects is still increasing rapidly and the wind power fleets are relatively young, some extra cost must be expected over time. According to a prediction based on statistics and analysis, the cost of global wind power O&M will increase from \$12 billion to more than \$27 billion in the next decade (Fig. 12).

In comparison with onshore wind farms, O&M costs of offshore projects are higher owing to higher cost of access to the location and maintenance of the towers and cabling. The marine environment is operationally more difficult than onshore, thus adding to the O&M costs. In Europe, the O&M costs were evaluated to be in the range from USD 109/kW/year to \$140/kW/year in 2017, which is expected to decrease to less than \$80/kW/year by 2025.

6 Load Factors[11]

The load factor of an installation is an important figure in the power industry where it quantifies how effectively installed assets are being used. In the wind power context, the load factor is calculated as the ratio of the amount of electrical energy actually produced by a wind project over a period of a month or a year to the total energy that could have been produced over the same period if the plant had operated continuously at its "nameplate" rated output for the same period. A given turbine achieves the nameplate output at a maximum operating wind speed, specified by the manufacturer, above which overspeed protection will be provided by active stall, braking or feathering. (Find chapter location for speed control.) The load factor falls below 100% due to winds below or above the maximum rated, and will be reduced further if the plant is not always electrically loaded to its maximum available instantaneous output power. However, this further reduction may not apply in some cases where the turbine is connected to a large grid that can be assumed to accept all, or almost all, of the turbine's instantaneous available output power at the instantaneous wind speed.

The average load factor of a wind system during its entire operating life is used to make certain assumptions, and any evaluation of the system cost must depend on these. In terms of the global level, the load factors increase consistently year by year as shown in Fig. 13. Clearly, the average load factor for onshore wind farms is between 25% to 30%, whereas that for offshore types is between 40% and 45%.

As an example, for the United Kingdom, a typical leading country in wind power applications, the load factors for onshore and offshore projects, averaged over UK installations, reached 32% and 43%, respectively, in 2016. There was only modest variation over the four UK countries (England, Scotland, Wales and Northern Ireland) and over the three years from 2014 to 2016. Offshore factors were always larger than onshore for the same country and year. A more detailed breakdown can be found in the listed literature.[13–15]

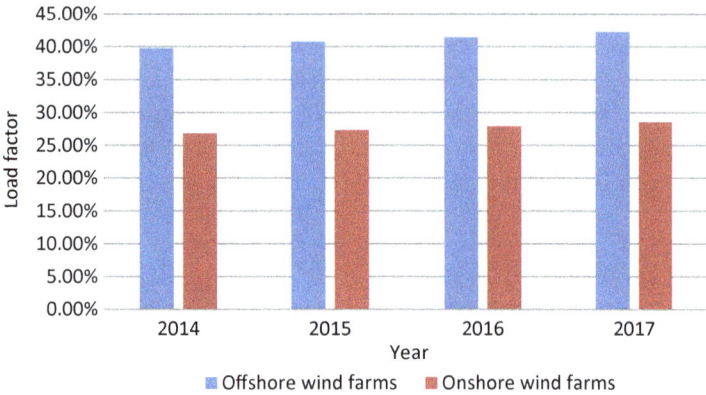

Fig. 13.   Global weighted average load factors for onshore and offshore wind plants.[12]

This reveals that the loading of offshore wind farms is systematically more efficient than for onshore ones, which is clearly because the offshore winds are more consistent. Offshore wind speeds are also higher on average.

7   Future Development of Wind Power Generation

Over the last two decades, tremendous advances have been achieved in the wind energy market, especially in installed energy capacity. This took place not only in Europe and North America, but also in China, India and other countries in Asia. Wind energy is certainly emerging as a serious competitor to more established but harmful energy sources such as gas and coal. Research also suggests that future wind energy will be the most cost-effective electrical power source compared to many others, including other renewables.

Besides mere expansion, the future development of the wind turbine industry will lie mainly in the technical development of wind turbines, and especially in the use of materials. There are still improvements to be made in improved system performance and efficiency, and lower component costs.

7.1   Turbine technology development

The main driver of turbine technology development is the return from increasing the turbine size, swept areas and hub height. The maximum size of rotors has gradually increased. Up to 2016, a typical operational offshore wind turbine was of 8 MW capacity with 164 m diameter blades. Figure 14 shows the average rotor diameter of wind turbines from 2015 to 2017.[16]

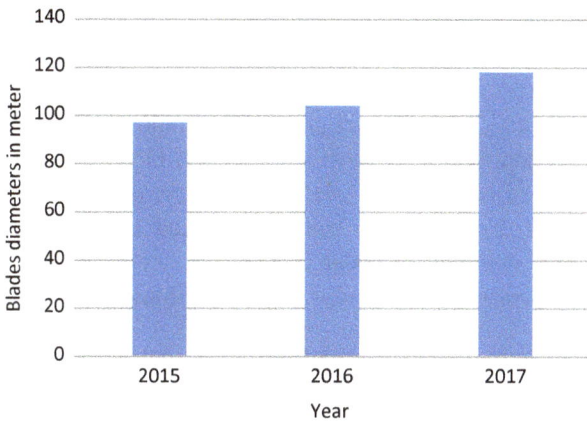

Fig. 14. Average rotor diameter of wind turbines from 2015 to 2017.[16]

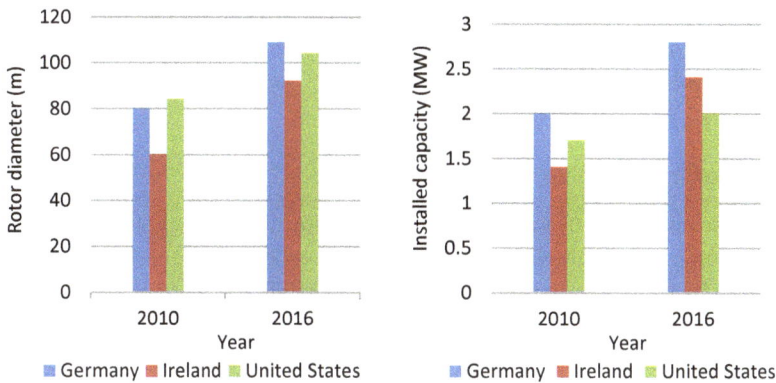

Fig. 15. The average rotor diameter and installed capacity evolution in three countries.[17]

Upward trends in rotor diameters and installation capacities in three countries, including Germany, Ireland and USA, in the years 2010 and 2016 are clearly shown in Fig. 15, and will continue worldwide.[17, 18] According to General Electric, the world's largest wind turbine will stand at more than 260 m from base to blade tip, with 220-m diameter rotors. With a 12 MW turbine, it can yield up to 67 GWh of gross annual energy, providing enough clean energy to power 16,000 EU households.[19] Such technology advancement has been mainly driven by Europe where there are space constraints and hence the need to exploit the marginal wind sites as much as possible.

Continued growth of turbine blade lengths implies in itself an increase in swept areas and therefore in power yields. It also requires greater hub-heights, which raise the output further by allowing the rotors to access higher wind speeds.[17] However, continued technology advancement in this direction will encounter challenges; a larger and heavier turbine requires a taller tower with a higher load bearing capacity, thus resulting in higher installation costs. Nevertheless, this is balanced by increased electricity yield, which may still lead to lower LCOE.[17]

With significantly longer blades, mechanical loads on turbines increase drastically, hence requiring different structural design for safe operation. This must be based not only on actual structural dynamic conditions under aerodynamic action, but also on a deep understanding of the environmental factors in which these multi-body rotating structures operate.[20] Transportation of long rotor blades becomes an issue. Though road upgrades are currently a more economical option, as the trend continues, the industry will eventually need to face the challenge of building segmented blades which can withstand the pressure exerted by the cyclo-stochastic patterns of the wind loading.[18–20]

7.2 *Material development*

Future development of wind energy generation technology will also depend on adopting new materials and improving production methods.[21] A crucial issue in the design and selection of materials for wind turbines is their fatigue properties. Most of the components in a turbine will have to endure 400 million fatigue stress cycles over its whole service lifespan.[22] This high cycle fatigue resistance is even more demanding than for aircraft, automotive engines, bridges and most other man-made structures. With the ever-increasing turbine size and length, gravitational loads and flapwise loads escalate, imposing even higher requirements on the structural stiffness. The stiffness-to-weight ratio becomes an important parameter for the materials and the cycle fatigue resistance also has to be increased further. Continued efforts have been made in developing new materials and structures which are stronger, lighter and can give high structural stiffness. In time, the materials will be expected to withstand high loading stress and reduce damage while extending the turbine operational lifetime[7] up to 20–25 years.

It is important to appreciate that fatigue resistance is at least as important as simple sheer strength. Where strength is effectively the stress (pressure) at which permanent deformation or failure occurs, the fatigue resistance is related also to toughness, which is defined as the strain (fractional elongation) as the material approaches failure. Non-intuitively, materials with higher toughness generally have higher strain, i.e. deform more, before failing, and this is related to the resistance to propagation of cracks. A material with poor toughness is also described as brittle, and glass itself is generally a strong but brittle material. This point is revisited in the following.

Wind turbine blades and nacelles are typically made from composite materials, and are the most costly part of a turbine. The generator machinery is mainly reliant on metals, while the tower can be metal or concrete (another composite material) reinforced by metal. Composites are composed of strong but discrete elements (usually fibers or sometimes particles), embedded in a continuous material called a "binder" or "matrix", and have long been known for being stronger than either component alone. Fibers may be laid down in a parallel or unidirectional (UD) pattern, or in various crossing patterns, as for example in a woven cloth.

By far, the most widely used composite for wind turbine blades is glass fiber, while carbon fibers have attracted increasing interest. Other fibers are aramid, polyethylene and cellulose, all of which have moderate mechanical properties, and low or very low densities.[22–24]

Glass fibers: these are made from materials such as silicate, colemanite (a borate), aluminum oxide and soda, and are the most commonly known and used among fibre-contributed composites. Glass fibres are manufactured by extruding melted glass produced from a specially designed furnace which has small holes at the base. On wind turbine blades, E-Glass and S-Glass types are used.[21] E-Glass has low cost, effective insulation and low water absorption rate. The stiffness of composites depends on the volume content and stiffness of fibers. In general, the stiffness, tensile strength and compressive strength of UD composites increase proportionally with the increase of fiber volume content. However, when the volume content of fibers approaches 65%, the fatigue strength of the composite will be diminished and "dry" areas without resin may appear among the fibers. In practice, glass occupies around 75% by weight of the glass/epoxy composites which are used in wind blades.[23]

New glass fibers can have substantially better performance, at greater cost, than E-glass fiber. For example, S-glass ("strong" glass, a non-borate glass chemically based on silica, alumina and magnesia) has tensile and flexural strengths 40% higher than those of E-glass; it also shows 10–20%

higher performance in compressive strength and flexural stiffness, but its price is around 10 times that of E-glass.[24] Windstrand[TM] is a trade name for several superior grades of glass fiber, around 15% stiffer and 30% stronger than E-glass.[25]

Carbon fibers at present offer the largest opportunity for alternatives to glass fibers. Compared with glass, they have much higher stiffness with lower density, and can be exploited in thinner, stiffer and lighter turbine blades. However, their performance in damage tolerance, compressive strength and ultimate strain is lower than for E-glass fibers and their price is much higher.[26, 27] Also, fiber misalignment and waviness have a big impact on carbon fiber reinforced composites. Even small misalignments can result in significant reduction of compressive and fatigue strength. At present, the Vestas (Aarhus, Denmark) and Siemens Gamesa (Zamudio, Spain) companies usually apply carbon fiber composites in building structural spar caps of large turbine blades.[28] (A spar is the longitudinal reinforcing element of an aircraft wing or turbine blade, and the caps are its outer edges.)

Other materials that can form strong fibers for use in composites are aramid (aromatic polyamide, e.g. Kevlar) polymers and minerals such as basalt. Aromatic polyamide fibers have the advantages of high mechanical strength, toughness and damage tolerance; the drawbacks, which must be recognized, are low compressive strength, low adhesion to polymer resins and moisture absorption.[28]

Basalt fibers are closer to the ideal and have been used in many small wind turbines with encouraging results. They demonstrate good mechanical characteristics, being 30% stronger, 15–20% stiffer and 8–10% lighter than E-glass fiber. They are cheaper than carbon fibers.[28] Hybrids obtained by mixing basalt and carbon fibers have been described.[29]

Hybrid Composites: These consist of two or more types of fibers in a common matrix. In recent years, there has been a lively interest in such materials, particularly those which are based on a combination of glass fibers with one of the stiffer, but more expensive, carbon fibers. Examples are E-glass/carbon, E-glass/aramid, and they have the potential to replace pure glass or carbon. An experiment by Ong and Tsai on an 8m turbine, using hybrids to completely replace pure carbon or glass reinforcements, showed that they would decrease the weight by 80%, but with a 150% increase in cost. With only partial (30%) replacement, the cost would increase 90%, while the weight would be reduced by 50%.[30, 31]

Several studies have considered the strength and damage mechanisms of hybrid composites, showing for example that incorporating glass fibers in carbon fiber reinforced composites can allow the improvement of their impact properties and the tensile strain-to-failure of the carbon fibers. Manders and Bader observed an enhancement of the failure *strain* of the carbon fiber reinforced phase when carbon fiber is combined with less-stiff higher-elongation glass fiber in a hybrid composite.[32] As mentioned above, this implies an increase in toughness. However, computations showed that the dependency of the composite *strength* on the glass/carbon ratio is V-shaped, with a minimum at the content of the order of 60% carbon, i.e. the hybrid strength can under some conditions be lower than the strength of both pure glass and pure carbon composites. This observation was confirmed experimentally. Thus, while the hybrid composites seem to be a very promising group of composites for wind energy, additional investigations are still required to find materials with definitively optimized performance compromises, and a conclusively optimal composition.

Matrix materials for Composites: The matrices of composites used in wind blades have mostly been composed of thermosets, including epoxies, polyesters and vinylesters. Thermoplastics have been used on a smaller scale. Thermosets occupied around 80% of the market of reinforced polymers. The advantages of these include the possibility of room or low-temperature cure and lower viscosity which eases infusion and improves processing speed. Polyester resins used in earlier composites were gradually replaced by epoxy resins in the development of large and extra-large turbines. Nevertheless, recent research argued for the return to unsaturated polyester resins; these have shorter cycle times and improved energy efficiency in production. This research also suggested newly developed polyesters could fulfil all the demands of strength and durability for large wind turbine blades. Matrix materials with fast curing at low temperatures remain an important research area.[33]

As a potential substitute for thermoset matrices, thermoplastic composites have the merit of recyclability, but they need a high processing temperature, which leads to higher energy consumption and which possibly influences the fiber properties. Moreover, another weakness of these composites is that they are hard to produce as large (over 2 m) and thick (over 5 mm) components because of much higher viscosity. In comparison to thermoplastic matrices whose melt viscosity is approximately 102–103 Pas, that of a thermosetting matrix is around 0.1–10 Pas. The melt temperature of thermoplastics is lower than their decomposition temperatures, and hence they can be reshaped during the melting process. However, fatigue behavior of thermoplastics is generally inferior to that of thermosets, no matter whether used with carbon or glass fibers, despite the fracture toughness of thermoplastics being better.[34, 35] Other possible advantages of thermoplastics are the possibility of automatic processing and the unlimited shelf lifespan of raw materials.[36]

8   Conclusions

The pressing worldwide requirement to reduce fossil fuel dependency implies that wind energy production will be expanding continuously in the foreseeable future, and high reliability and extended lifetime of wind turbines are the important pre-conditions for supporting such expansion. The future in wind turbine technology presently seems to lie in the installation of large and extra-large turbines, to be placed in large, dedicated wind parks, mostly offshore. These future turbines will be made from even more advanced composites that are lightweight, highly durable, fatigue resistant and damage tolerant — and nevertheless cheaper to manufacture.

References

1. W. Shepherd and L. Zhang (2017). *Electricity generation using wind power.* Singapore: World Scientific Publishing Co. Pte Ltd.
2. "Wind Farm Expenditure", *Irena.org*, 2018. https://www.irena.org/docume ntdownloads/publications/re_technologies_cost_analysis-wind_power.pdf.
3. https://www.nrel.gov/docs/fy18osti/70363.pdf.
4. http://www.irena.org//media/Files/IRENA/Agency/Publication/2018/Ja n/IRENA_2017_Power_Costs_2018.pdf.
5. https://www.irena.org/documentdownloads/publications/re_technologies_c ost_analysis-wind_power.pdf.
6. https://www.statista.com/statistics/499491/us-wind-turbine-price-index/.
7. http://www.irena.org/-/media/Files/IRENA/Agency/Publication/2016/IR ENA_Power_to_Change_2016.pdf.
8. https://www.statista.com/statistics/506774/weighted-average-installed-cos t-for-onshore-wind-power-worldwide/.
9. https://www.nrel.gov/docs/fy17osti/66861.pdf.
10. https://www.statista.com/statistics/506756/weighted-average-installed-cos t-for-offshore-wind-power-worldwide/.
11. https://www.statista.com/statistics/555654/wind-electricity-load-factor-uk/.
12. https://www.gov.uk/government/collections/energy-trends#2018.
13. https://www.gov.uk/government/uploads/system/uploads/attachment_dat a/file/462357/Regional_renewable_electricity_2014.pdf.

14. https://www.gov%2Cuk/uploads/system/uploads/attachment.data/file/55
6271/Renewable_electricity.pdf.

15. https://assets.publishing.service.gov.uk/government/uploads/system/uploa
ds/attachment_data/file/647335/Regional_renewable_electricity_2016.pdf.

16. https://www.statista.com/statistics/263901/changes-in-the-size-of-wind-tu
rbines/.

17. Adnan, Z. Amin, Renewable Power Generation Costs in 2017, International
Renewable Energy Agency (IRENA), ISBN978-92-9260-040-2.

18. C. Draxl (2012). On the predictability of hub height winds. Roskilde: DTU
Wind Energy.

19. https://www.ge.com/renewableenergy/wind-energy.

20. A. Velazquez Hernandez (2014). Model updating and structural health monitoring of horizontal axis wind turbines via advanced spinning finite elements
and stochastic subspace identification methods.

21. http://citeseerx.ist.psu.edu/viewdoc/download?doi=10.1.1.464.5842&rep=r
ep1&type=pdf.

22. B. Eker, A. Akdogan, and A. Vardar (2006). Using of composite material in
wind turbine blades. *Applied Journal of Applied Sciences*, **6**(14), 2917–2921.

23. L. Mishnaevsky and P. Brøndsted (2009). Statistical modelling of compres-
sion and fatigue damage of unidirectional fiber reinforced composites. *Composites Science and Technology*, Elsevier.

24. P. Brøndsted, H. Lilholt, and A. Lystrup (2005). Composite materials
for wind power turbine blades. *Annual Review of Materials Research*, **35**,
505–538, doi:10.1146/annurev.matsci.35.100303.110641.

25. L. Mishnaevsky, *et al.*, (2017). Materials for wind turbine blades: An
overview. *Materials*, **10**, 1285, doi:10.3390/ma10111285.

26. C. Red (2008). Composites World. Viewed on 05. 11 2011 from Wind turbine
blades: Big and getting bigger. https://www.compositesworld.com/articles/
wind-turbine-blades-big-and-getting-bigger

27. D. Fecko (2006). High strength glass reinforcements still being discovered.
Materials Today, Elsevier Ltd, **50**(4), 40–44.

28. https://infogram.com/carbon-fiber-vs-fiberglass.

29. https://www.compositesworld.com⟩articles⟩basalt-fibersalternative-to-gl.

30. https://www.build-on-prince.com⟩basalt-fiber.

31. L. Mishnaevsky Jr. and G. Dai (2014). Hybrid and hierarchical nano reinforced polymer composites: Computational modelling of structure–properties
relationships. *Composite Structures*, Elsevier, **117**, 156–168.

32. L. P. Mikkelsen and L. Mishnaevsky, Jr. (2017). Computational modelling
of materials for wind turbine blades: Selected DTU wind energy activities.
Journal of Materials (Basel), **10**(11), 1278.

33. P, W. Manders and M. G. Bader (1981). The strength of hybrid glass/carbon
fibre composites, Part 2 A statistical mode. *Journal of Material Science*,
16(8), 2246–2256.

34. S. Tusavul, R. Fragoudakis, A. Saigal, and M. A. Zimmerman (2014). Thermoplastic materials for wind turbine blade design. *The World Congress in
Civil, Environmental and Material Rresearch, Busan Korea*.

35. R. Orozco (1999). Effect toughened matrix resins on composite materials for wind turbine blades, Montana State University-Bozeman, MSc Degree Report.
36. R. T. D. Prabhakaran, T. L. Andersen, Jakob Ilsted Bech and H. Lilholt, Thermoplastic composites for future wind turbine blades — Pros and Cons, 18. *The International Conference On Composite Materials.*

Chapter 4

Geothermal Energy

Magnus Gehringer

Consent Energy,
Washington, DC, USA and Reykjavik, Iceland
magnus@consentenergy.com

Abstract

Geothermal energy is natural heat from the Earth used to produce electricity or for direct heating of buildings, greenhouses, swimming pools, aquaculture ponds and for industrial processes. Geothermal resources have been identified in nearly 90 countries and utilized to produce electricity in 24 countries and for direct heating in 72 countries. The worldwide installed capacity for electricity generation is approximately 14 GWe and for direct heating is approximately 16 GWt. Generally, geothermal projects require a high initial capital investment, but low annual operating costs. The common types of power plants are high temperature flash steam and low/medium temperature binary cycle types. Due to its high availability factor, geothermal power provides the baseload. Large high-temperature projects provide competitively priced electricity. The greatest risk in developing a geothermal project is in locating, defining and developing the resource resulting in the drilling of a number of production and re-injection wells to produce the necessary hot water and/or steam. Most geothermal resources are found along tectonic plate boundaries and volcanic regions where high-temperature resources are near the surface. Generally, environmental impacts are small and can be mitigated.

1 Introduction to Geothermal Energy and its Utilization

In the core of planet Earth, radioactive decay constantly generates heat that slowly moves to the outer layers, including the mantle and the surface.

This process is found everywhere and is reflected in an average global temperature gradient of 30°C/km. Therefore, in most areas of the planet, one can find temperatures of 90°C at 3 km depth, just to give an example. In areas with volcanic activity, however, temperature gradients of over 100°C/km can be found.

This heat energy is called "Geothermal Energy". Since the radioactive processes in the core of the Earth are continuous, it can be classified as a renewable energy source. Theoretically, geothermal energy could provide mankind with all its energy needs many times over, but in practice it has been shown to be difficult to develop geothermal energy projects due to two main reasons:

- **Resource risk:** Our knowledge of the thermal characteristics of the subsurface is quite limited, though geophysics and geology have made progress over recent decades. Even after a completed state-of-the art exploration program, the actual existence of a geothermal reservoir and its potential can only be confirmed by test drillings, followed by months of well testing. These test drillings require an investment in the range of about US$ 10 to 25 million, depending on the site in question, and this investment is at risk if the resource cannot be confirmed or the potential turns out to be insufficient for the intended purpose.
- **Project finance and profit margins:** Investors in geothermal projects need to have access to "patient" capital. The project cycle is on average seven to nine years until utility scale power generation (>25 MWe) can start and revenues can be generated. For the first phases including test drilling, three years are a typical estimate These facts and the fact of the resource risk mentioned above, make it very difficult, if not impossible, for private sector companies to develop a geothermal power project from the beginning. Furthermore, it is common knowledge that geothermal power generation cannot offer the high returns that investors are targeting. Especially in countries where power tariffs are low, profit margins will be too low to generate interest from the private sector.

Geothermal fields are generally located around volcanically active areas often located close to boundaries of the tectonic plates. Figure 1 provides a world map with the main plate boundaries showing the most important geothermal areas.[1]

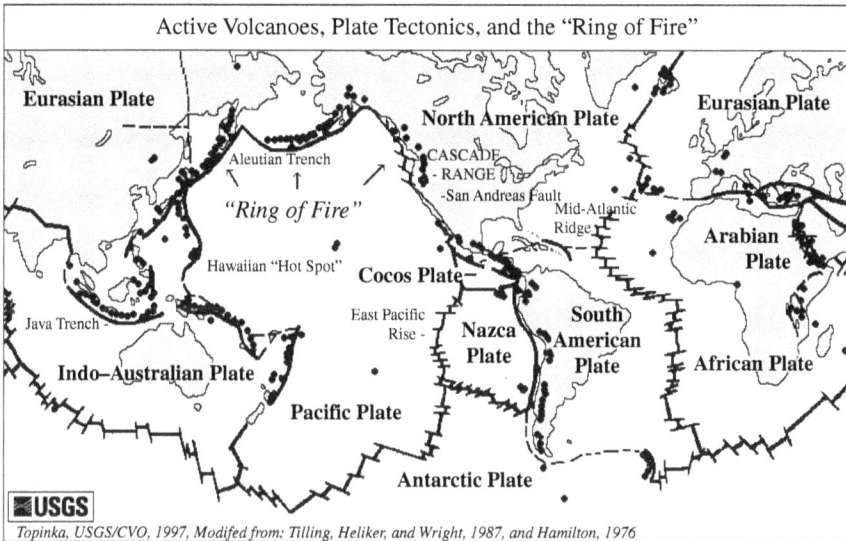

Fig. 1. World map of plate boundaries.

Geothermal energy has been used since the early 20th century for electricity generation and even longer for direct uses.[a] The installed capacity of geothermal power generation has increased from about 2 to around 14 Gigawatts electric[b] (GWe), and direct heat use has increased from about 2 to around 16 GWth (2011) during the last four decades.

2 World Overview of Utilization

Geothermal resources have been identified in nearly 90 countries and there are quantified records of geothermal utilization in approximately 72 countries. As of 2010, electricity is produced by geothermal energy in 24 countries. With around 50%, Kenya has the highest share of geothermal power in its energy mix, followed by Iceland and El Salvador with 25% of their electrical power generated from geothermal resources. The United States has

---

[a]The term "direct use" refers to applications other than power generation, e.g. space heating, industrial process, greenhouses, cooling, etc.

[b]In generating electricity, the output power is often denoted as Megawatt electric (MWe) or Gigawatt electric (GWe) as opposed to the input power in the form of heat which is designated as Megawatt thermal (MWth) or Gigawatt thermal (GWth).

Table 1.   Installed Geothermal Capacity of 12 Leading Countries.[2]

Country	Capacity (MWe) Oct. 2018	Share of National Power Generation	Rank
USA	3591	0.3	1
Indonesia	1948	3.7	2
Philippines	1868	27	3
Turkey	1200	0.3	4
New Zealand	1005	14.5	5
Mexico	951	3	6
Italy	944	1.5	7
Iceland	755	30	8
Kenya	676	51	9
Japan	542	0.1	10
Costa Rica	207	14	11
El Salvador	204	25	12

the most installed geothermal capacity, 3,500 Megawatts electric (MWe), followed by Indonesia and the Philippines with around 1,900 MWe each.

The 12 countries having more than 200 MWe of geothermal power capacity installed are outlined in Table 1. In total, about 38 countries are considered to possess significant geothermal potential that could be developed to augment, and in some developing countries increase, their current power generating capacities and make them less dependent on energy imports.

2.1   Projected future growth of geothermal power generation

Global geothermal power generation has increased steadily over the years, but slower than most other forms of renewable energy. Between 1990 and 2010, the globally installed capacity increased from 5.5 to nearly 11 GWe. This translates into 100% growth over 20 years or an average of 5% increase per year. As shown in Fig. 2, future growth is expected to be more rapid and 17.5 GWe are projected to be installed globally by 2020 and 24.5 GWe by 2030. These figures represent geothermal power from hydrothermal resources only.

The additional capacity in the coming years can be expected to come from the current front runners (USA, Mexico, New Zealand, Japan and Iceland), as well as from some European countries (e.g. Italy, Greece, the Balkan countries), Turkey and its eastern neighbors, and several Middle

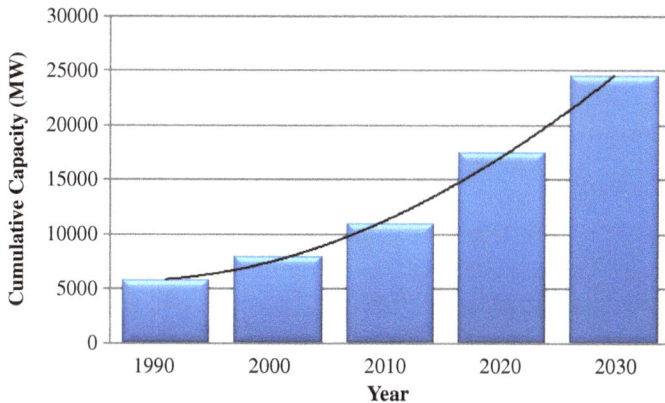

Fig. 2.   Forecast of global installed geothermal capacity 1990–2030.[3]

Eastern countries, such as Yemen. Australia and some South Pacific island states also may be able to utilize their hydrothermal resources for power generation.

Looking beyond 2030, significant additions in installed capacity also can be expected in the following countries and regions:

- **Pacific Asia:** Malaysia, Papua New Guinea.
- **Africa:** Tanzania, Eritrea, Sudan, Somalia, Malawi, Zambia, Burundi, Rwanda, Uganda, Democratic Republic of Congo, Mozambique, Madagascar, Comoros Islands and several North African countries.
- **Latin America:** Guatemala, Honduras, Panama, Colombia, Ecuador, Bolivia and several Caribbean island states, including Cuba and Haiti.[c]

3   Geothermal Geology

Figure 3 shows the components of a typical hydrothermal (steam or water based) volcanic-related geothermal system, which are, from bottom to top:

- The magmatic intrusion, also called hot body, where hot magma intrudes exceptionally far into the Earth's crust, often caused by movements of the continental plates.
- The actual geothermal reservoir where steam and/or hot water are trapped under a tight, non-permeable layer of rocks (the cap rock) and heated by the hot body below.

---

[c]Text and graph based on forecast in Geothermal Handbook (2012), see reference 3.

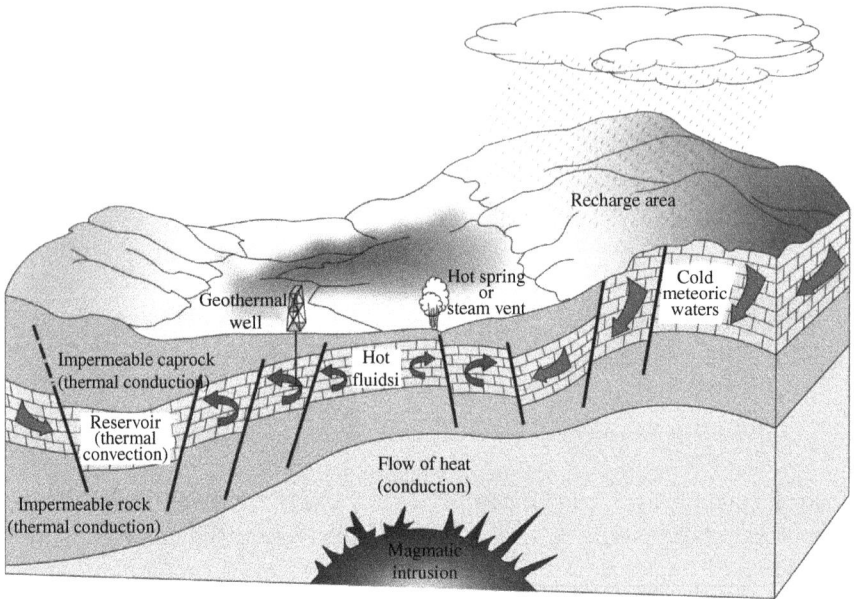

Fig. 3.   Schematic view of an ideal geothermal system.[4]

- Water/precipitation coming from recharge areas like lakes, rivers or the seas, which provide cold waters slowly seeping down through the ground to lower layers through cracks and along faults in the rocks.
- The geothermal wells which tap into the geothermal reservoir and access the hot steam or water, which is then transferred from the wells through pipelines to the power plant or for direct use application, from where the spent fluids are usually returned into the reservoir.

4   Development of Geothermal Power Generation Projects

Geothermal electrical generation projects have seven key phases of project development before actual operation and maintenance (O&M) commences. According to the schedule shown in Fig. 4, it takes approximately seven years to develop a typical full-size geothermal project from a completely undeveloped site. However, depending on the relevant country's institutional and regulatory framework, geological conditions, location and availability of financing, the project development time could either be reduced or prolonged by several years.

Each phase of geothermal project development consists of several tasks. After each milestone, the relevant developer — either a project company or

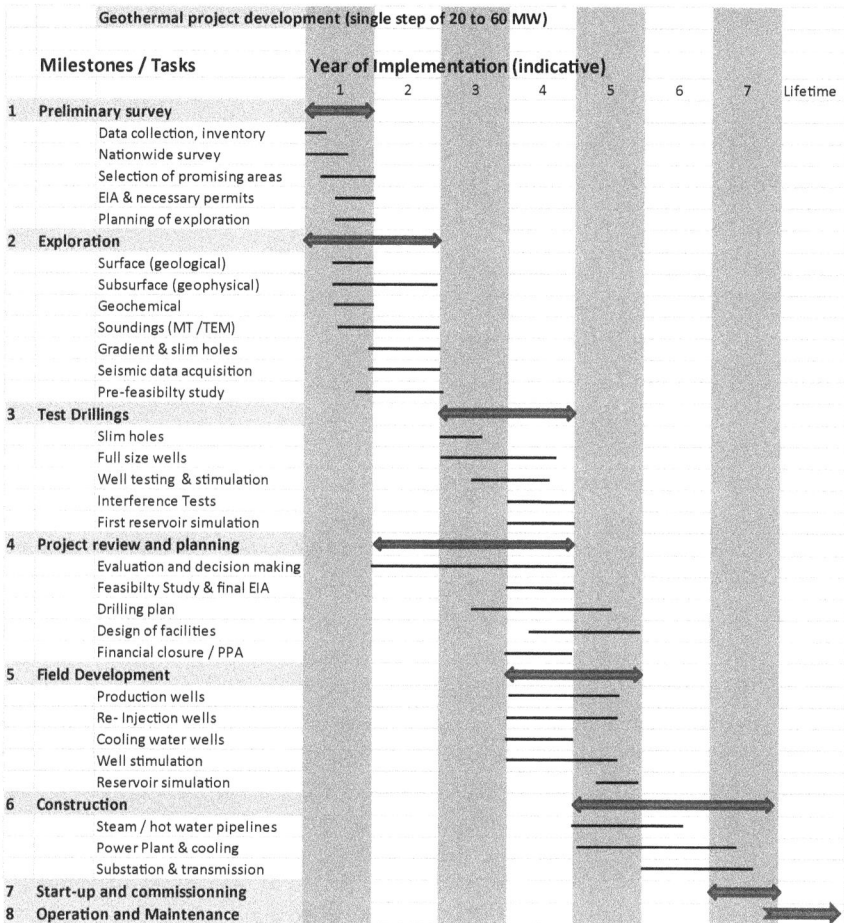

Fig. 4. Geothermal project development schedule for a unit of approx. 50 MWe.

a country's institution — will have to decide whether to continue developing the project or not and whether to assume the risks of the next phase. The first three phases, or milestones, take the developer from early reconnaissance to actual site-specific scientific research through test drillings. This first part of the project development (which could be broadly called the exploration stage) either confirms the existence of a geothermal reservoir suitable for economically viable power generation or not; therefore, it is usually seen as the riskiest part of project development and therefore also the most difficult to finance. If the results from the first three phases, including the test drillings, are positive and the geothermal potential is confirmed, phase 4 is initiated with the actual design of the power project, including the feasibility study, engineering design and financial closure. Phases 5–7 comprise the development of the project itself, consisting of the drilling of geothermal production and re-injection wells, construction of pipelines, the power plant and its connection to the transmission system. The following paragraphs describe the development phases in more detail and provide indicative figures for a cost analysis.

4.1 *Exploration*

Phase 1, the preliminary survey, includes a first reconnaissance of a geothermal area based on a nationwide or regional study. If no geothermal master plan studies are available, developers usually conduct their own studies based upon available literature and data, or execute their own reconnaissance work to select the areas to apply for exploration concessions. Methods for this purpose include geological ground studies as well as satellite and airborne research. Phase 1 is important to establish the rationale and assess the need for the project in question and at the same time to find a justification to enter into investments induced by the exploration and test drillings (Phases 2 and 3). Costs for this first phase can be estimated as US$ 0.5 million up to 1.5 million, which shows that they can vary considerably according to the data available, the documentation required and the size and accessibility of the area being considered for geothermal power generation. Phase 1 usually takes from several months up to one year to complete.

Phase 2, the exploration phase,[d] consists of detailed geophysical surveys and may include the drilling of temperature gradient or core holes to further confirm the preliminary resource assessment and starts as soon as the project developer is satisfied with the results of Phase 1. In total, the second phase can take up to two years, depending on the size and accessibility of the geothermal field and the data already available.

Costs for the activities under Phase 2 can be significant, from 5 to 10% (indicative figures) of the total project costs, depending on the project size. Costs for conducting Transient Electromagnetic and Magnetic Telluric

---

[d]In this context, the term "exploration phase" refers to the second phase in the detailed breakdown of the project cycle. This usage is distinct from "exploration" in a broad sense, which consists of the first three phases including the test drilling phase. The latter usage is more common for the oil and gas industry.

resistivity measurements, seismic surveys, gravity measurements or drilling gradient or core holes depend on the accessibility of the geothermal site and the availability of tools, equipment and the capacity of staff to operate the equipment and interpret the results. While minimum exploration costs for a geothermal site would in many cases be US$1 to 2 million, every single gradient well could add US$0.5 to $1 million to that figure. Since all geothermal fields and projects are different, it is difficult to generalize the required investment costs for Phases 1 and 2.

Phase 3, the test drilling phase, is the last of the exploratory phases. At the end of this phase the project developer should be able to decide, based on scientific evidence and the characteristics of the reservoir, whether or not the project warrants being continued, i.e. will it be economically viable to build and operate a power plant, or should the project be abandoned?

In the beginning of this phase, a drilling program is designed to target and to confirm the existence, the exact location and the potential of the reservoir. Usually a set of 3 to 5 full-size geothermal wells[e] are drilled. In some locations due to accessibility issues, and the availability of such things as water and electricity at the geothermal field, it may be necessary to start with the drilling of slim holes (i.e. holes with a smaller diameter) that can be drilled with lighter equipment (drilling rigs) than full-size wells, and at lower cost. Slim holes drilled to penetrate the reservoir will provide much needed information relative to depth, temperature and may allow for some level of well testing to determine the potential for productivity. Cost for a three-well reservoir confirmation drilling program (full-sized production wells) could run or even exceed 20 million US dollars depending upon location, accessibility, depth and completion diameter. Upon completion of the drilling program, a long-term flow test as well as injection tests should be conducted to determine the potential for interference between production and injection wells. The completion of a reservoir engineering report at this point will be critical to determining the conversion technology to be employed, the optimum size of the power plant and as a prerequisite for the obtaining of mezzanine financing that will allow for further field development, completion of feasibility studies and the start of engineering design.

---

[e]For example, a full size well could be 1.5–3.5 km deep and have a bottom hole diameter of 7–8 inches. The top (surface) diameter can be over 20 inches.

4.2 *Drilling and well testing*

Despite the fact that a discovery well or wells have been successfully completed and tested during Phase 3, risks associated with drilling out the well field (production and re-injection wells) to support the planned power generation facility are still considerable and should not be taken lightly. In fact, even in well-developed and producing fields, more than 20% of all wells drilled are unsuccessful. The outcome of the drilling of a well can be defined over the entire spectrum from success to failure. Success is usually defined as a well capable of an energy output suitable to meet the intended utilization. A successful well would have the following attributes:

- High flow rate
- High enthalpy of the geothermal fluid
- Low non-condensable gas content
- Low potential for scaling and
- Low potential for corrosion.

Unfortunately, even if only one of the above is unfavorable, the well may be a failure for the purpose of meeting the intended use. An important consideration is the fact that the first wells drilled in a new field tend to be less productive than those drilled at later stages of well field development due to increased understanding of the subsurface and of what drilling techniques are best suited in the particular geological environment.

Well output can vary considerably from field to field and even from well to well. The worldwide average output per well is just over 4.0 MWe per well. However, some fields have been successfully developed when output is as low as 2.0+ MWe per well and in highly unusual circumstances, average well output can exceed 7–8 MWe with some wells producing in excess of 40 MWe.

It is thus critically important that as much information be gained from the drilling of each well as possible in order to continually improve drilling success as well as output per well. Every effort should be made to incorporate such information as quickly as possible into the drilling plan as it has a major impact upon overall project economic viability.

Critical components of the drilling plan include rig specifications, casing set points, the cementing program, maximum angle of deviation (especially critical in the upper most sections of wells drilled for binary plants in order to ensure the ability to set pumps), angles of deviation to intersect faults and fractures at optimum angles, program for overpressure and/or underpressure drilling, the mud program, drill bit selection, completion type, logging program, etc. Other critical considerations include blow out prevention and contingencies for bringing a well back under control if a problem should occur.

4.2.1 *Well testing*

Well testing is the process used to obtain the data needed to fully evaluate the performance of a geothermal reservoir including long-term productivity potential and the potential for well field interference between production and injection wells and resulting in premature temperature declines.

Much of the needed information critical to any well test is obtainable during drilling operations and thus diligently gathering and recording such data is critical to future interpretation of test data. Such data includes indications of the depth and nature of production intervals, lost circulation zones, temperature anomalies, rapid changes in drilling rates, evidence of fractures and/or alteration minerals in drill cuttings.

The first test to be conducted upon completion of drilling operations is a rig test. It is of limited duration (several hours to 1–2 days) conducted while the drilling rig is still on the hole and is primarily to clean out the hole. to provide initial estimates of well productivity and allow for the sampling of reservoir fluids for chemical analyses. The test data is of critical importance for making decisions regarding further drilling and completion of the well and provides the basis for planning future tests of the well.

Following the completion of the well, a single-well controlled flow production test of several days to a week or more is run. The single-well test provides information relative to the well's productivity and allows for temperature and pressure surveys to be run under stabilized and controlled conditions. Pressure and/or water level draw down and build-up data allows for the making of initial estimates of reservoir parameters.

Longer term production and interference tests of +/− 30 days duration are generally run to determine reservoir parameters, to detect boundaries or recharge sources, to evaluate scaling and corrosion potential, to assess reservoir capacity and to identify any factors that could result in a need to modify the design of the utilization system. In interference tests, pressure or water level changes in observation well are monitored while producing from the production well. Injection tests serve the same purpose as do production tests, i.e. to determine well performance, to locate injection zones and determine various reservoir parameters. It must be stressed that reservoir properties obtained from injection tests are not necessarily applicable to the same reservoir undergoing production testing. This is due primarily to the thermal dependence of such parameters as fluid density, formation porosity and fracture aperture.

It may be required to inject during a production test due to the need to dispose of the fluids produced and if this is the case, careful planning should allow the production and injection tests to be run concurrently.

Of primary concern in long-term reservoir productivity and optimization planning is the potential for premature thermal breakthrough resulting in temperature degradation of the production fluids and resulting in decreased power output. It is thus important to include tracer testing in any comprehensive well-testing program. Results of tracer tests can help in the siting of injection wells and in the management of production and injection.

4.3 *Mitigation of the resource risk*

A comprehensive strategy for minimizing resource risk exposure should consist of the following approaches:

1. **Portfolio exploration**, in which the country to some extent explores and evaluates multiple geothermal fields simultaneously, thereby increasing the probability of finding at least one viable site and reducing the chance of overlooking significant development opportunities.
2. **Parallel development** of the fields selected from the portfolio to multiply the pace of development and to reduce time and costs.
3. **Stepwise expansion**, reducing the risk of reservoir depletion and pressure drops by developing a geothermal power project in cautious increments/steps, determined by reservoir data. As a rule of thumb, a pilot power plant (e.g. wellhead generator with 2 to 10 MWe capacity) should be installed to gain solid geophysical data about the reservoir over a period of 1 to 2 years. Thereafter and based on this information, a utility scale power plant can be built in incremental steps of e.g. 25 or 50 MWe, depending on the field potential and pressure-drop measured.
4. **The country's minimum system demand for electricity** will determine the maximum amount of geothermal capacity that can be installed, since geothermal is base load and steam is difficult to control, geothermal power plants usually do not follow load.

Unlike the oil and gas industry that serves the global market, demand for geothermal power is localized and limited by a specific country's/ region's minimum system demand (base load). This means that the entire demand for geothermal power may be met by a relatively small number of productive geothermal fields.

4.4 *Determination of power plant size by demand analysis*

Two factors strongly determine the highest possible installed capacity, and thereby power generation, of a geothermal power plant: (i) The demand for electricity in the country or within the system, and (ii) The potential of the geothermal reservoir.

The electric load within a country depends on power generation on one hand and on demand for power on the other. The system only functions optimally if generation and demand are nearly the same at all times.

The Fig. 5 presents a typical country load curve, in this case with two peaks over the day. Depending on the country, load curves have different shapes, according to the system demand they reflect. Demand may also vary considerably from season to season as well as over the course of a day.

It is a common and justifiable view that geothermal power plants cannot be adapted to following system demand, because the power plant depends on hot steam coming from the ground, which is a steady source and somewhat difficult to control except in the case of binary systems where the wells sometimes need to be pumped. Therefore, geothermal power plants are usually deployed to provide base load to the system,

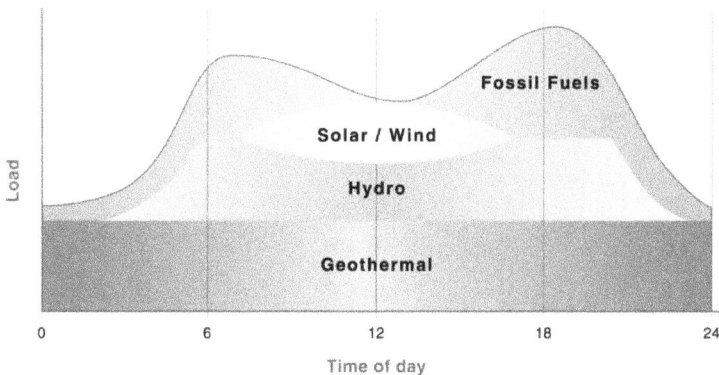

Fig. 5. Simplified load curve with typical fuel sources.[3]

as shown in the Fig. 5. Other power sources can adapt more quickly to the demand; this applies for example to diesel generators and hydropower plants. Along with gas power plants, these sources can be used to track the load within the system. Dispatching various power sources depends on whether they can be used for base load, for peaking operations and how fast they can adapt to changes in the system demand. Another argument against operating a geothermal plant in a load following or dispatchable mode is the high up-front capital cost and low fuel cost as well as the operation and maintenance (O&M) costs, all of which favor base-load operation. It is common practice to grant priority dispatch to geothermal power as well as to most other forms of renewable energy, in order to decrease the use of fossil fuels and make the water for hydropower in its reservoirs available for use over a longer period of the year. For this reason, the size of a geothermal power plant should not exceed the minimum system demand.

5 Geothermal Power Generation Technologies

The standard power plant classification defines 5 different types: binary, single flash, double flash, back pressure and dry steam. The relative share in power generation (2010) of each of these technologies is reflected in Fig. 6.

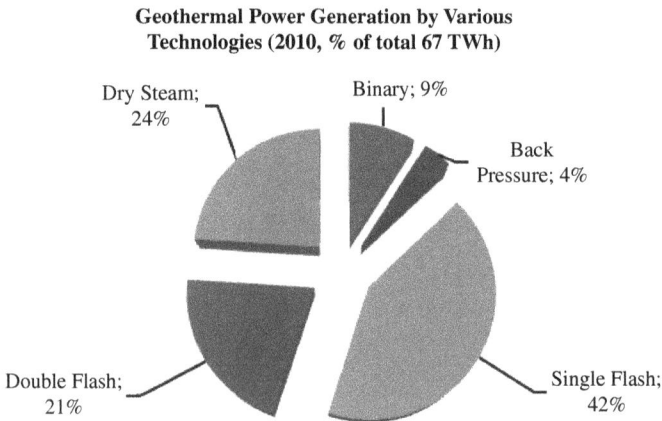

Geothermal Power Generation by Various Technologies (2010, % of total 67 TWh)

Dry Steam; 24%

Binary; 9%

Back Pressure; 4%

Single Flash; 42%

Double Flash; 21%

Fig. 6. Relative share of various conversion technologies, 2010.[3]

Source: Adaptation from Bertani, 2010

Large-scale electricity generation mainly takes place through the use of conventional steam (single or double flash) or binary power plants, depending on the thermal, pressure and chemical characteristics of the geothermal resource.

5.1 *Flash plants, Condensing Units*

Flash plants, also called Condensing Units or Conventional Steam Cycles, usually come in sizes from 25 to 60 MWe per turbine and as one pressure stage application (single flash) or with two pressure stages (double flash). Obviously, the double flash version is significantly more efficient but also costlier. The decision as to whether or not a double flash cycle is worth the extra cost and complexity can only be made after a thorough economic evaluation of the project.

In Fig. 7, the two-phased fluids (liquid-vapor mixture) enter through a separator. While the separated steam continues to the turbine to be used for power generation, the fluids (brine) either get re-injected right away, or could be used for secondary uses. The steam is expanded though a turbine and then cooled in the condenser (often cooling towers) so it can be re-injected into the reservoir (only fluids can be re-injected since steam (gas) cannot economically be pumped into the reservoir). Secondary uses include bottoming cycles, i.e. additional power generation with usually a binary unit, or other "waste heat" uses as discussed below.

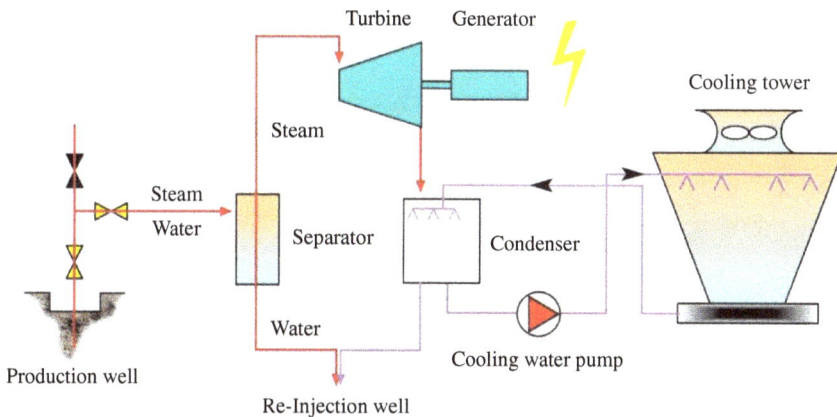

Fig. 7. Concept of flash/condensing geothermal power plant.[3]

5.2 *Binary cycles*

Binary cycles, also called Organic Rankine Cycles or ORC, are being built
to generate power from fluid temperatures below 100°C. For reasons of
project economics, 120°C is seen as the absolute minimum fluid temperature to deploy a binary cycle. Binary cycles operate quite efficiently on
fluid temperatures over 140°C and up to 200°C.

As shown in the diagram below, a binary cycle has two loops: An open
loop for the geothermal fluids coming up from the reservoir, heat being
extracted by heat exchangers (evaporator or boiler) and the cooled geothermal fluids being re-injected into the reservoir. A second, closed loop is filled
with an organic working fluid (alcohol, usually iso-pentane) that has a very
low boiling point and creates steam to drive the turbine. This enables the
binary cycle to utilize such low fluid temperatures. After exhaustion of the
steam in the turbine, the working fluid has to be cooled down (re-liquefied)
in a condenser, so that it can be pumped back into the storage tank and
finally be used again (see Fig. 8).

Binary cycles are being produced by many suppliers, with different
working fluids and come in sizes from 100 kWe to 20 MWe. They are often
used as "bottoming units", i.e. the binary plant uses the (waste) brine from
the separator of a flash plant and the residual fluids (also called waste heat)
coming from the turbine. Combined, the two energy sources can provide
10–20% of additional power to any power project and therefore bottoming units should always be part of the initial project/power plant design.

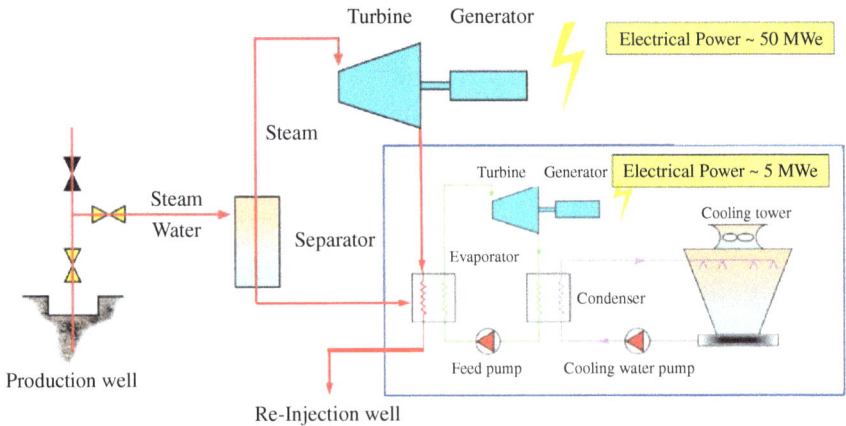

Fig. 8.   Concept of typical binary power plant, ORC or Kalina.[3]

This is a logical step to make best possible use of the energy from the reservoir. Binary cycles are often more cost-intensive per MWe installed but are usually easy to maintain and operate. In addition, and contrary to flash plants, binary cycles usually do not allow greenhouse gas emissions into the atmosphere.

5.3 *Additional technologies*

Dry steam: In a few areas of the world, namely in California (USA), Italy, Indonesia and to a lesser extent in Japan and New Zealand, geothermal reservoirs can be found that produce pure hot steam due to the low reservoir permeability. These resources can be utilized by a condensing power plant, as introduced above, but in this case there is no need for a separator to separate fluids and steam. Generally, dry steam units are large (>50 MWe) and operate with high efficiency.

Back pressure units: Back pressure units are steam turbines that exhaust the incoming steam, whether dry or wet, directly into the atmosphere. This makes them compact, simple to install and run and the cheapest choice available. However, they are usually only used for a limited amount of time, e.g. as test units or wellhead generators, until a more appropriate solution can be implemented. Back pressure units have a relatively low efficiency as compared to the other technologies mentioned earlier — generally the electrical output is about 50% of the other technologies, which means they generate significantly less power out of the same amount of steam, and they can, depending on the chemical composition of the fluids/ steam, be hazardous to the environment as they lack equipment to deal with non-condensable gasses and in addition generate a great deal of noise pollution.

5.4 *Efficiency and power generation costs of the technologies*

Flash power plants are used to generate power from high enthalpy resources. The efficiency for the overall conversion process of heat energy into electricity may be around 20% and power generation costs for utility scale power projects of over 25 MWe are usually between US$ 0.04 and 0.11 per kWh.

For binary cycles, efficiency is less mainly due to the heat exchange for the second, closed loop. With an overall efficiency of 10–14%, the power generation costs of binary units are significantly higher than from flash power plants. Since binary units are usually also smaller in size than flash plants, an estimate for production costs from binary would be between US$ 0.08 and 0.2 per kWh, depending mainly on resource temperature and plant size.

5.5   Direct uses and utilization of waste heat

5.5.1   Direct use of geothermal energy

In low enthalpy geothermal fields with fluid temperature of 60–120°C, direct use may be a feasible option. This includes the same uses as mentioned above for waste heat use, but, due to low fluid temperatures, does not include power generation. Therefore, the same uses are available for direct uses as for the use of waste heat, which is discussed in the following paragraphs. In addition, a later section presents a case study on the utilization of waste heat.

5.6   Options of waste heat utilization

What makes geothermal different from all other renewable energies is its availability factor of 90% or higher for new and modern power plants. With usually over 8,000 hours per year of operating time, geothermal power plants provide reliable and renewable base-load power. Additionally, once a geothermal power plant is operational, it can also be used in multiple ways to enhance the projects overall economic result. This is called multiple use, cascaded use or utilization of residual — or waste — heat. Furthermore, the acronym CHP (combined heat and power) has become increasingly popular.

Figure 9 is an idealized diagram showing cascaded use of geothermal energy, based on the example of a small (2 MWe) binary power plant in Iceland, which, located as far as 18 km from its wells, uses the residual heat of the fluid (after power generation) for nearby industries (food industry), for domestic heating for the entire town, for fish farming and finally for snow melting in streets. As a result, the energy contained in the fluids is nearly completely used. In addition, geothermal power plants can also be connected to industries that produce a lot of heat, such as steel mills, biomass power plants or waste incinerators. Their produced heat can be used to enhance the temperature of the geothermal fluid in order to increase power production.

The options for multiple uses of the energy, and the fact that small modular units with up to 5 MWe capacity are readily available and easy to

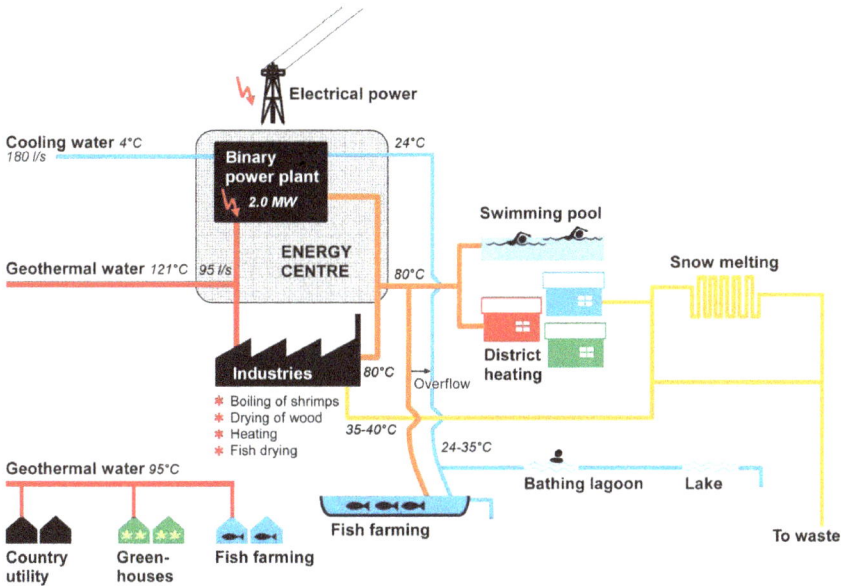

Fig. 9. Idealized diagram showing multiple uses of geothermal energy.[5]

install and operate, make geothermal power generation also a feasible option for smaller installations in remote and even off-grid locations, especially when they replace existing and more costly fossil fuel generation.

In general, the revenue stream from the use of the residual heat (waste heat) can improve the overall financial viability of both small size and industry scale (over 25 MWe) power projects, with additional revenue coming from:

- Bottoming cycles, i.e. usually binary units utilizing waste heat of >130°C from the turbine of a flash power plant. Bottoming cycles usually add around or over 10% to the power generation of a flash plant.
- Sale of flowers, plants or vegetables grown in greenhouses. Not only heat, but also CO_2 can be extracted from geothermal fluids.
- Canning factory for vegetables, fruits (e.g. jam, juices) or fish, meat, dairy products.
- Fish or shellfish farming, or other aquaculture products.
- Dehydration (drying) of fruits, nuts and other food products.
- Desalination of seawater to provide drinking water.
- Cold storages or freezing plants.

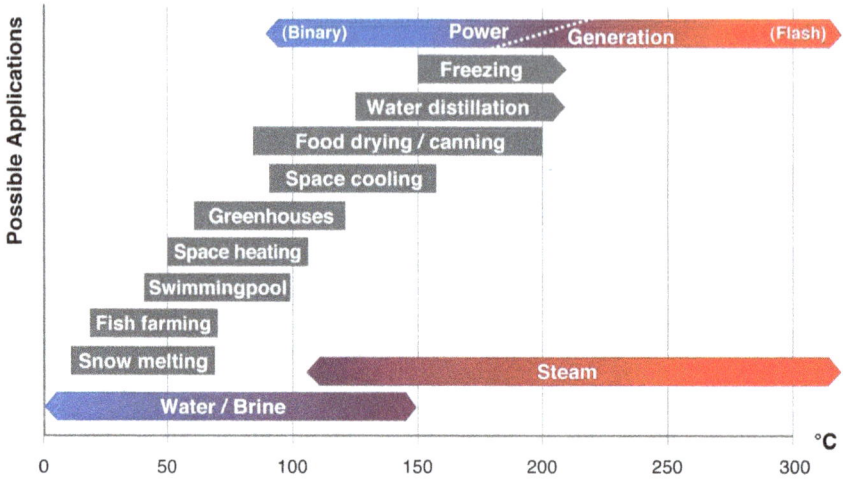

Fig. 10.   Modified Lindal diagram showing applications for geothermal fluids.[3]

- Use of waste heat for industrial processes, e.g. chemical, biological and food industry.
- Sale of hot water for district heating or district cooling purposes.
- Extraction of valuable minerals and salts from the geothermal fluids, e.g. silica, manganese, zinc and lithium. As a gas, CO_2 can be used for industrial purposes, e.g. the soft drink industry.

These options are site-dependent; some geothermal sites will offer the use of several of these options simultaneously, while others offer none. A more comprehensive overview of how geothermal fluids and steam can be utilized over their temperature range is given in Fig. 10.

6   Economics of Power Generation

Table 2 presents an indicative cost analysis for geothermal project development of a typical 50 MWe project in a geothermal field with no initial development and with wells of around 2 km in depth.[3] Power plants of 25 to 50 MWe can often be a suitable first step unit, which can be expanded or multiplied at a later stage, or used as a final unit. Wells generally range in depth from between 1.5 and 3 km, with an international average of around 2 km, which will be used in the following calculations. The cost figures include all exploration and drilling costs, as well as estimated

Table 2. Indicative Costs for Geothermal Development (50 MWe Generator Capacity) in Million US$.[3]

	Phase/Activity	Low Estimate	Medium Estimate	High Estimate
1	Preliminary survey, permits, market analysis[a]	1	2	5
2	Exploration[b]	2	3	4
3	Test drillings, well testing, reservoir evaluation[c]	11	18	30
4	Feasibility study, project planning, funding, contracts, insurances, etc.[d]	5	7	10
5	Drillings (prod and inject) (20 boreholes)[e]	45	70	100
6	Construction (power plant, cooling, infrastructure, etc.)[f] and cooling options (water or air).	65	75	95
	Steam gathering system and substation, connection to grid (transmission)[g]	10	16	22
7	Start-up and commissioning[h]	3	5	8
	TOTAL	142	196	274
	In million US$ per MWe installed	**2.8**	**3.9**	**5.5**

Notes: [a]Costs for survey depend heavily on size and accessibility of area. Costs for EIA depend on country regulations.

[b]Depending on methods used and accessibility and size of area.

[c]For 3 to 5 drillings with variable depths and diameter, from slim hole to full size production wells.

[d]Studies and contracts provided by external suppliers or own company. Conditions and regulations of relevant country.

[e]Depending on depth, diameter and fluid chemistry, casings and wellhead requirements in terms of pressure and steel material/coating. Also influenced by underground and fractures (drilling difficulty and time).

[f]Power plant prices vary by system used and supplier, but most impact comes from infrastructure (roads, etc.)

[g]Depending on distance from plant to transmission grid access point, and on distance between boreholes and power plant.

[h]Standard industrial process. Power plant may need fine-tuning for some time and minor adaptations. For high estimate, major changes, repairs and improvements are needed to supply power according to PPA.

financing costs for the development of a hydrothermal reservoir for power generation.

Given this variability of geothermal investment costs, a useful question to ask is: how high can the investment cost of geothermal become before it ceases to be economically competitive? This can be accomplished by comparing geothermal with other base-load technologies, such as steam turbines running on heavy fuel oil (HFO) or coal, medium-speed diesels (MSD) on HFO, and eventually large hydropower plants.[6]

Based on an oil price of around US$75 per barrel, the economic break-even investment costs for geothermal would be:

- US$8,900 per kWe installed, as compared to steam turbines operating on HFO (Heavy Fuel Oil).
- US$7,000 per kWe as compared to MSD (Medium Speed Diesel).
- US$5,200 per kWe as compared to steam turbines using coal.
- US$4,400 per kWe as compared to large hydropower facilities with capacity factor 60%.

Geothermal energy is an unusual case, in that the main source of uncertainty lies in the investment cost. The only comparable case is hydro, where investment costs may vary according to how geological characteristics develop during construction; however, a priori determination of expected costs can be gauged with some accuracy, whereas in the case of geothermal, the actual exploration cost is a major factor in the economics of a potential project. Computer models which accurately take into account this source of uncertainty to quantify the tradeoffs with competing resources have yet to be developed.

6.1 *Financing geothermal power projects*

Figure 11 gives a generic overview of project risks and investment required, as well as project development phases and financing options. The high risk factor and financial requirements of the test drilling for geothermal resources makes public sector support for geothermal projects nearly inevitable. Typically, for the fear of the project not being financially viable, the private sector will be extremely hesitant to finance a geothermal project until the geothermal resource is proven. Especially the test drilling phase has been a bottleneck to development; there are very few options for developers to finance their projects through this stage. Some countries can afford offering grants and other support mechanism, like geothermal funds, to push geothermal projects over this hurdle.

The impact of financing options and related costs can be shown quantitatively by a financial model based on cash-flow analysis. For a typical 50 MWe geothermal power project, the model rendered a levelized cost of energy (LCOE) of US$ 0.05 to 0.07 per kWh and, most importantly, shows the enormous impact of public support in early project stages for the overall

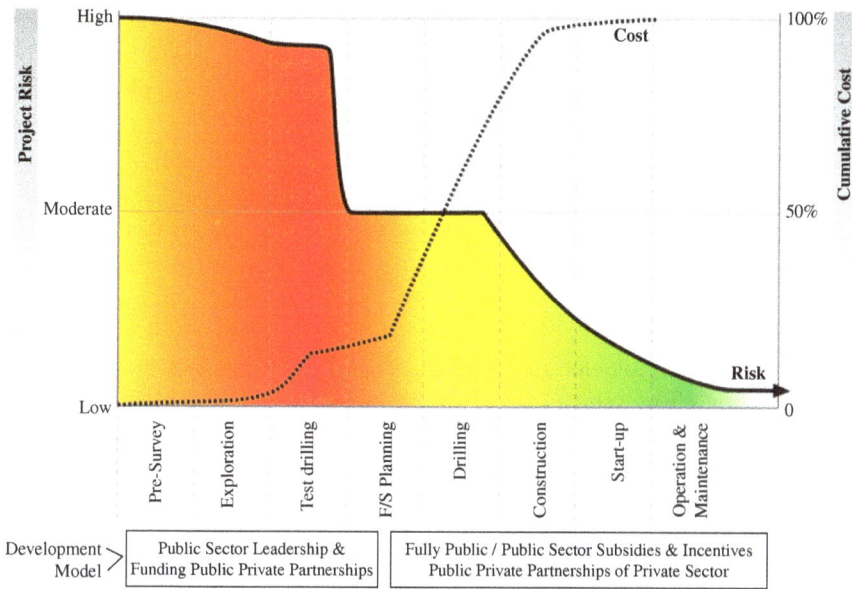

Fig. 11. Financing a typical medium size geothermal power project (50 MWe).[7]

financial viability of the project. In short, financing costs encountered by private sector companies financing their project with venture capital can easily double the LCOE/required power tariff, thereby rendering the project uneconomical.

6.2 *Incremental costs*

Several countries with geothermal resources have other energy resources which can be used for power generation at a lower cost. Examples are Canada for hydropower and Indonesia for coal, to name only two. The incremental costs of geothermal power generation can present a barrier to the deployment of geothermal power plants in these countries, especially in those cases where the power price paid by the end consumer is low. In the example of Indonesia, coal power plants situated within the coal mine area usually generate electricity at a lower cost than geothermal. In addition, sometimes public utilities do not want or are not allowed by the government to raise consumer tariffs.

Incremental costs can also result from the location of a geothermal site and the necessary investment for new transmission lines. Typically, areas promising a viable geothermal resource will not coincide with electric load centers. Cities with large populations are not generally built on active geological faults or on the flanks of active volcanoes that are likely to be able to support large-scale geothermal power plants (i.e. flash plants). This introduces the additional risk of finding geothermal reservoirs of sufficient size to build power plants large enough to justify the cost of transmission lines to the load center. Small geothermal binary plants which take advantage of lower geothermal temperatures found at greater distances from primary geothermal areas may have application when closer to load centers.

To summarize, even in countries richly endowed with geothermal resources suitable for electricity production, geothermal power generation is clearly not always the cheapest option. In spite of its usually attractive power generation costs of 5 to 11 US$ cents per kWh on average and a capacity factor of at least 80%, the economic attractiveness of geothermal power projects need to be carefully assessed compared to other competing sources of electricity (e.g. hydropower, gas, coal, etc.) available in countries producing exportable fossil fuels.

6.3 *Alternative design of geothermal funding mechanisms*

The preceding sections mention the importance of public financial support during the initial project phases, or until the potential of the reservoir and the commercial viability of the power project are proven. Besides several small regional geothermal funding and support schemes that provide grants, often on a political or bilateral basis, two kinds of Risk Mitigation Funds (RMFs) are most commonly deployed:

1. Insurance policies covering test wells; and
2. Combinations of grants and loans for geothermal exploration and test wells, available by application on a case-by-case basis.

Both approaches have been used for decades, but neither has had unbridled success. Loan and grant funding have not substantially accelerated geothermal development, while the insurance scheme has not promoted a single megawatt of geothermal power on the ground. As one representative example, the IFC had a fund of US$10 M to subsidize insurance premiums for Turkish geothermal projects in cooperation with Munich Re insurance, but gave up after five years without having approved a single application from private-sector developers. Similarly, the African Rift Geothermal Development Facility (ARGeo), in its original World Bank setup, had funds of around US$11 M to be used as grants for geothermal exploration and drillings in five East African countries; but in 2010 the World Bank canceled the fund and reallocated the remaining funds for other uses. The reasons for these programs' lack of success are discussed in the following.

The insurance scheme (e.g. InterAmerican Development Bank, IDB in Mexico) is based on subsidizing the insurance premium for one or more test wells for a project developer. The developer can be a private-sector company or a public utility or power company. The rationale behind the insurance scheme is that with test drillings de-risked, developers should be able to receive financing for drilling costs from commercial banks.

However, while the drilling of five test wells might cost around US$20–25 M, a developer would still be required to provide an equity share of 30% (i.e. US$6–$8 M) to receive debt financing. Additionally, the insurance scheme unfortunately does not go far enough in mitigating some of the other important risks, such as evaluation of the power-generation potential of a reservoir. While test drilling might confirm steam availability, it does not ascertain or confirm the long-term production capability of the reservoir and, hence, the investor/developer is still unsure about whether to invest long-term funding in the project. Often, the financial status of many geothermal developers is not sufficiently robust to meet the typical costs encountered during the development of a project, such as maintaining highly qualified staff, constructing the necessary infrastructure and funding overhead for operations and maintenance. Unlike in oil and gas operations, where viable resources can more easily be exploited to generate immediate revenue, geothermal developers do not receive any revenue until a resource's steam is converted to sellable electricity. Over the two or more years' period of test drillings, developer costs, including company infrastructure and overhead, can be overwhelming. Of course, insurance helps (by guaranteeing that a developer will get back the money spent on drilling), but basically the developer must bear all other costs. This means that after two to three years of operations, a developer might only get a refund of the drilling costs. In the end, this would be a huge net loss for the developer. Thus, the insurance scheme is not a viable business model; obviously, businesses need to make profits. In other words, in this same time frame, a developer would very much prefer to develop two wind or solar projects and make profits on each one. Beyond profits, such successful projects would improve a developer's reputation, while spending two years on drilling and then getting refunded by an insurance company would not look so impressive. Hence, in order to be successful, the insurance scheme would have to provide a business model by which developers could pay for their staff, overheads and other costs — plus make a profit. From an ethical standpoint, this would be very difficult to justify.

The **grant and loan support scheme** (e.g. the Geothermal Risk Mitigation Facility, GRMF, based in Ethiopia and serving all of East Africa) has been used for decades. In this scheme, applications for grants and loans are received by a donor-funded agency, but success in actual installed megawatts developed on the ground is usually less than expected. One reason for this is that the potential of a geothermal reservoir needs to be scientifically proven through long-term well testing before developers can commit to a project and before banks can in turn commit to funding the developers. A grant for exploration might increase developer understanding about the likelihood of steam in the subsurface, but it will not *prove* reservoir potential in terms of megawatts that can be developed. In more detail, the reasons for slow success of the grant and loan support scheme can be summarized as follows:

1. Government geothermal agencies that apply for grants, loans or any form of assistance from a Risk Mitigation Facility (RMF) scheme have limited capability and capacity to plan, manage, and implement their projects; and, as a result, overall implementation progress is slow. In these cases, it has been recommended that such agencies employ the services of a company with the prerequisite expertise to assist it in the implementation of the project under the RMF scheme.

2. Private companies applying to this kind of RMF are sometimes not qualified and do not fulfill the three basic conditions for suitable developers of having: (a) a track record of previous projects; (b) in-house capacity and expertise; and (c) the capital assets to securely implement their project. For example, if a developer is awarded grant funding for 40% of their exploration costs, they will not receive these funds until they can prove that they have secured the remaining 60% of the total costs. This has been shown to be a major bottleneck for many IPPs, and has slowed the development process.

3. The availability of grants and cheap loans attracts private companies that under normal circumstances would not have tried to enter the geothermal business. These companies are aware they do not fulfill the three prerequisites for developers and often are speculators trying to

use the grants to do initial development of the field and then sell it to another entity that can actually develop the project. This speculation slows down geothermal development and inflates costs, and is only to the benefit of the speculators. In some countries, this behavior of private-sector developers has led to license-hoarding.

4. Similarly to the insurance scheme, in the RMF scheme the problem might occur that large, powerful and experienced companies cannot apply for funding because they cannot be sure about their business model. Large companies that would fulfill all three prerequisites for developers and could actually develop the project on their balance sheet do not apply for these grants and loans. Their problem is that they cannot be sure about reservoir potential. Due to their large size and higher overhead costs, they are not interested in small, e.g. 5 MWe, projects, but have targets on the scale of hundreds of megawatts as a minimum size for power-generation projects. As long as they have no long-term proof of reservoir potential, these companies will be very reluctant to enter into the geothermal business. They can better plan their business and profits in other renewables or industries.

This explains why sufficient drilling, long-term well testing and resource evaluation with a pilot power plant are needed to provide interested parties with sufficient scientific data about the megawatt potential of a geothermal reservoir to incentivize investment. Currently, existing RMFs do not provide this kind of funding or risk mitigation, which explains these schemes' failure to facilitate geothermal development. Well testing and resource evaluation over a time frame of one to two years can usually only be accomplished by having several active production and re-injection wells operating with a power plant or other heat removal process. The power plant provides cash flow, but also cools down the fluids from the production wells so that they can be re-injected.

6.4 *New approach for an effective RMF structure*

Confirmation of medium/long-term production capability of a geothermal reservoir would require one to two years' production data from the reservoir, which can be achieved by including in the funding the construction of a small power plant of about 5–10 MWe. For the example of a 5 MWe power plant in a high enthalpy field, two to three active full-size wells are required. By so doing, the technical risks in further development of the field are reduced. No new technologies should be used; all technologies and equipment should be proven and off-shelf. To ensure success in implementation, the RMF would probably hire a consultant who would supervise all project phases and use the portfolio approach of developing more or less identical projects in several countries. This would provide a steep and efficient learning curve, while allowing for optimal economies of scale. Again, the technical risks in further development of the field would be reduced. Once the pilot power plant has been operational for a year or two, scientific data about the reservoir and its subsurface location, potential and fluid chemistry will have been collected and evaluated. Then, a smaller (as well as large-scale) developer could make an informed decision about how to develop and finance the project's expansion.

A feasibility study for each geothermal field, consisting of technical, financial and environmental (ESIA) studies will provide scientific facts about the reservoir potential and commercial viability to IPPs and/or publicly owned power companies. If a state-owned agency had developed the first phases of a geothermal project with support from a RMF, that agency or entity could now make an informed decision about the price tag to put on the project for international bidding or other kinds of sale/transfer to the private sector. At this stage, the concession has a clearly defined value, and a sale price in the form of a concession fee can be demanded by the developing entity. Since the field has been largely de-risked, the buyer and future developer of the field will be willing to pay a fair price consisting of actual development costs plus a risk premium. In addition, the reduced risk means that the parties could produce bankable project documentation to obtain funding through commercial banks and funding institutions as if for a construction project. Since all geological, technical, commercial and financial data would be readily available for further project planning, *overall* risks are reduced and further expansion of the projects might be of interest to both the public and the private sector. In addition, the RMF could decide to sell the de-risked and ready-to-fund field to the future developer and thereby become a rotating fund. The RMF can, however, decide not to recoup any of the investments made in the project and give the project as a grant to the future developer.

What makes this RMF structure unique is mainly the deployment of a small geothermal pilot power plant that needs to be operated for at least one year, with all its data collected and scientifically interpreted. The conceptual model would be refined as data becomes available. It is also very important that all raw data acquisition and processing should be up to international best practices, so potential future developers would not doubt their quality or dispute the value of the field.

Existing RMF structures have either failed or had only limited impacts on the pace of geothermal project development. At the same time, the structures of these funds are complicated and multilayered, such that their operation costs significant amounts of donor money. Initial ideas for a new and more effective funding and support facility include a streamlined internal structure, reduced drilling costs through project coordination, reuse of the same pilot plant for several power projects and finally the option of recouping some of the support funds (i.e. the rotating fund scheme).

6.4.1 *Features of an effective alternative RMF*

1. Builds in-country geothermal capacity (manpower and institutional) and a resource database by focusing on cooperation with the public sector (usually, a country's dedicated geothermal agency). The access point for private-sector companies is at a later stage, when the resource potential and commercial viability of a reservoir have been proven by long-term operation.
2. Provides grant support for exploration and test drilling. At least two active wells (one production and one re-injection) will be needed to retrieve appropriate long-term and interference data from the reservoir. These data should prove the viability of installing a small pilot plant to utilize the resource for power generation.
3. Makes grant funding available for Environmental and Social Impact Assessment (ESIA) and pre-feasibility studies, looking at market issues, tariffs, off-takers, transmission and other issues that might determine and influence project viability at a later stage.
4. Provides a country's dedicated geothermal agency with debt financing for a modular pilot plant (e.g. ~5 MWe capacity) to be connected to the two or more test wells that have been drilled. The power from this plant can be used on- or off-grid. Experienced technical consultants support the project development, while the country's agency participates in planning, installation and operation of the project. The pilot wellhead power plant generates power over a timeframe of at least one whole year. Data received from this operation, along with volumetric assessments and simulations, will reveal significant amounts of data about the reservoir leading to a firm understanding of the conceptual model

of the project, reservoir interference and, most importantly, the overall sustainable potential of the reservoir. At this stage, the availability of steam for an initial power plant (see our discussion on the Stepwise Approach mentioned earlier) can be guaranteed and, at the same time, a strong indication of the future total capacity of the steam field will be obtained. This will allow future developers to make informed decisions about whether to bid for the field.

5. Allows the pilot power plant to be moved and the process repeated at the next developed geothermal field, after the initial field has been fully tested and gets expanded.

6. Produces a feasibility study for each geothermal field (consisting of technical, financial and environmental components) to provide scientific facts about the reservoir potential and commercial viability to IPPs and/or state-owned power companies. These parties can now make informed decisions about further developing the field, or they might choose to put the project out for bidding by tendering out the geothermal field. Since the resource risk has been removed, the parties can use their bankable project documentation to obtain funding through commercial banks and funding institutions as if for a construction project. Furthermore, the bidding parties should be supplied with a full package of all licenses and permits needed to move directly into financing the first phase of the power project (e.g. a 25–50 MWe power unit).

7. Provides consultancy and supervision to the project developer through all project phases, to ensure that the feasibility study results in project financing and to guide the developer through any legal, institutional or technical problems that could arise.

6.4.2   Fund requirements

It is estimated that the fund requirements for confirming the potential of one geothermal field per country (including exploration data analysis to develop a database and the drilling of three to four wells of 2000 m depth) would be around US$15–25 M for test drilling; while for one or two 5 MWe pilot power plants, approximately US$12 million would be required. Assuming that about four to five countries would be initially ranked for development, it seems that the initial grant funding could reach about US$40–45 M, while the RMF should have a total endowment of around US$100–150 M. The capital of the RMF would very much depend on how and to what extent, if at all, the fund would recoup its investments from successful projects. As outlined above, if the fund is organized such that developers pay a fair price for the de-risked and proven geothermal fields, the RMF could function as a rotating fund and would require less replenishment over time than conventional RMFs.[8]

6.5 *Evaluation of geothermal power projects*

Whether a private-sector company (IPP) considers developing a geothermal project or investing in one, or whether a donor/investor, a commercial bank or Multilateral Developing Bank is considering supporting a project or financing it, there are always several basic questions to be asked to evaluate the financial and technical parameters of a geothermal project. Here are a few of these questions:

(1) For IPPs the first question will be: Who are the previous IPPs operating geothermal power projects in the country and how successful have they been in their business?

(2) For the investor or bank, the first question will be: Can the developer prove that he has received the geothermal concession or permit, or at least that he will get it, or has a chance to get it? What other permits are available, or will be needed, and how difficult and costly will it be to obtain those?

(3) Does the country have a clear and supportive legal and regulatory framework, allowing projects to be developed by IPPs?

(4) How is geothermal data managed by the geological survey of the country and what data is available/accessible?

(5) For a certain project, what exploration has been done so far and what are the results?

(6) Has the availability of a geothermal resource been confirmed by drilling? If yes, were several test wells drilled, interference test done and wells tested for several months, to render reliable data about the potential?

(7) Who will be the off-taker for the generated power and what is his financial standing?

(8) Is a PPA with the off-taker available or what is required to get one signed?

(9) Does the off-taker or government offer a fixed tariff for geothermal power? Is it high enough to sustain the operations?

(10) Is the developer aware of the minimum system demand of the country and transmission capacity of the grid? Does he have access to all

relevant data about the system? Does he understand the necessity of having access to reliable transmission and distribution systems?

(11) Is the project only focusing on power generation, or does it promote a more holistic approach including waste heat use for secondary industries, which could multiply the project's benefit for the country and especially the region?

(12) Has the project been planned with mitigation of resource risks in mind? After test drilling, will the developer first deploy a wellhead unit or small pilot plant, and then increase capacity in small increments? The potential of the reservoir should be scientifically confirmed, by long-term tested wells, before designing a full-size power plant.

7   Environmental and Social Issues

Geothermal energy has several characteristics that make it appealing for power generation. Geothermal power plants provide base-load power with a high capacity factor; modern geothermal power plants can have a capacity factor of 90% or higher. They are also an ideal complement to hydroelectric power whose load-following capability allows a power system to serve peak loads. Once a geothermal power plant is up and running, there is little need for fuel, which contributes to low operation and maintenance costs. The multiple uses of geothermal resources, including for power generation, industrial heat, tourism and agricultural production, can enhance the economics of geothermal projects.

Some of the drawbacks of geothermal energy are associated with characteristics of the resource itself. Field depletion is a risk which can be mitigated by designing the geothermal development carefully with the stepwise approach. Depending on characteristics of the field, additional wells (make-up wells) may have to be drilled every few years (at a significant cost) to sustain the production rate. Additionally, there may be temperature drops of the steam being extracted, which can impair the ability to deliver the rated capacity of the power plant.[9]

The potential environmental and social impacts of geothermal plants compare favorably to fossil fuel technologies as well as other types of renewable technologies. Land use for a geothermal power plant and related resettlements are limited compared to other technologies. Impacts on air quality through plumes and smells can in most cases be mitigated. Land subsidence has occurred in cases when reservoirs have been overdeveloped, but it can usually be mitigated by re-injecting the geothermal fluids back into the reservoir to maintain the pressure necessary to sustain the production of steam.

8   Case Study: Use of Waste Heat from Geothermal Power Plants

From the above, it seems obvious that waste heat from geothermal power plants can be used for various kinds of food production and food processing industries. Outside of the capital city area, most countries seem to be focused on farming and food production. If waste heat from geothermal power plants could open new job opportunities, facilitate the creation of new products or enable keeping products fresher and more valuable through better storage options, it might become a tempting vision to try to utilize the waste heat from all geothermal projects. Needless to say, the heat from low and medium temperature geothermal fields ($<150°C$) can be used for the same purposes as waste heat from power plants. This is called direct use.

There are many ways to improve the economic and financial viability of a geothermal power project by using the waste heat from the power plant. Besides increasing profits from operations and providing additional tax revenues, these activities in many cases also save operation costs for the power plants by reducing the amount of fluids that have to be cooled down by cooling towers. In other words, the power plant operator also benefits from utilizing the energy from the hot waste fluids, because he then saves the costs for cooling the fluids down by cooling towers, at significant cost. Therefore, all stakeholders seem to benefit from using waste heat energy. However, it must be ensured that the waste fluids stay within the geothermal field, since they will have to be re-injected into the reservoir to stabilize the reservoir pressure in the subsurface and avoid reservoir depletion.

In total, the use of waste heat should improve the overall project benefits for the entire country as well as for the people in the region by the creation of additional jobs and the provision of new opportunities. All a power plant operator will have to do is to make the waste heat from the plant available to waste heat users, e.g. a developer of a canning factory, instead of cooling it down and re-injecting it right away. This requires two interfaces for the waste heat users; one, where the waste heat is received from the power plant, and another one where the fluids are returned to the power plant operator after the heat has been extracted. This process reduces the usage of cooling towers, but for security reasons, the cooling towers still have to be installed and be instantly operational. For this additional work, including providing the two interfaces, the power operator will have to be compensated well enough to ensure his firm and continued interest in providing his waste heat at all times to connected waste heat users.

8.1   *Examples of waste heat uses from a 50 MWe flash plant*

The following graph shows, without the deployment of a bottoming cycle, that a geothermal flash plant with a capacity of 50 MWe net produces around 100 MWth thermal energy (see Fig. 12).

How this amount (100 MWth) of thermal energy could be used will be shown by the following examples. In many countries, more than 50% of all produce (vegetables and fruits) rot before they reach the market. In fact, the shortage of ways to preserve food limits significantly the options for farming in these countries. Many crops, especially vegetables and fruits, are not grown because the transport to the next market is just too time-consuming and the production per farmer too little to justify direct transport.

To prevent the food from being spoiled before and during containment, a number of methods are used: pasteurization, boiling (and other applications of high temperature over a period of time), refrigeration, freezing, drying and vacuum treatment, to name a few.

The waste heat can theoretically be transported over quite a distance, e.g. 10 km or more, before it reaches the user of the heat energy. However, in order to save water, the transport medium for the heat, a closed loop cycle might be needed. Since the pipelines are costly, it might be worth

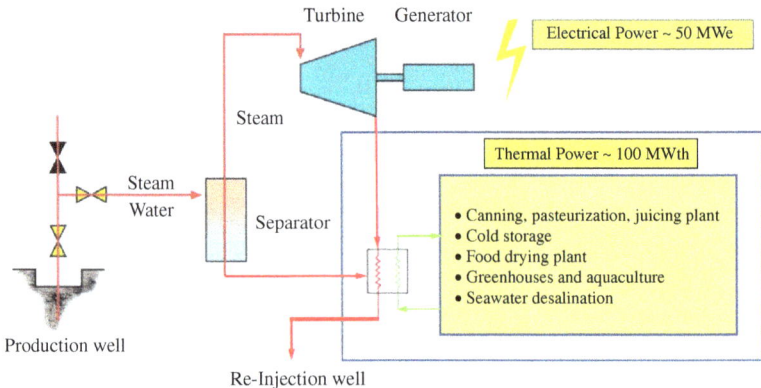

Fig. 12.   Energy flow to waste heat users.[10]

trying to locate the facilities using the waste heat as close to the power plant as possible.

8.1.1 *Canning and pasteurization*

Canning is a method of preserving food in which the food contents are processed and sealed in an airtight container. Canning provides a shelf life typically ranging from one to five years, although under specific circumstances it can be much longer. A freeze-dried canned product, such as canned dried lentils, could last as long as 30 years in an edible state. From a public safety point of view, foods with low acidity (a pH more than 4.6) need sterilization under high temperature (116–130°C). To achieve temperatures above the boiling point requires the use of a pressure canner. Foods that must be pressure canned include most vegetables, meat, seafood, poultry and dairy products. The only foods that may be safely canned in an ordinary boiling water bath are highly acidic ones with a pH below 4.6 such as fruits, pickled vegetables or other foods to which acidic additives have been added.

The available energy of 100 MWth from the example of a 50 MWe power plant would allow the operation of a canning company with a capacity of

— 200 tons per hour of canned products
— 100 cargo shipping containers per day could be filled and shipped.

Canning would likely have seasonal peaks and lows, and therefore not create equal revenues over an entire year. Depending on the sales products and based on a sales price of $0.5 to $1 per kilogram (two cans), a production of 4,500 tons per day might generate revenues from US$2 to $4million per day.

8.1.2 *Cold storage*

Refrigeration has had a large impact on industry, lifestyle, agriculture and settlement patterns. Refrigeration technology has rapidly evolved in the last century, while the increase in food sources has led to a larger concentration of agricultural sales coming from a smaller percentage of existing farms. Farms today have a much larger output per person in comparison to the late 1800s. This has resulted in new food sources becoming available to entire populations, which has had a large impact on the nutrition of societies.

Dairy products are constantly in need of refrigeration, and it was only discovered in the past few decades that eggs needed to be refrigerated during

Fig. 13.   Cold storage of Northland Inc., Wisconsin USA, size 25,000 m^2.

shipment rather than waiting to be refrigerated after arrival at the grocery store. Meats, poultry and fish all must be kept in climate-controlled environments before being sold. Refrigeration also helps keep fruits and vegetables edible longer (see Fig. 13).

The waste heat from a 50 MWe flash power plant has been estimated as 100 MWth. This amount of heat energy is sufficient to operate an absorption chiller system that could provide freezing temperatures of $-20°C$ to a cold storage (freezing plant) of 100,000 m^2. Compared to Fig. 11, this would be four of the cold storages shown above. With an estimated height of 10 meters, there would be plenty of room for cooled and frozen products for farmers and other industries from the entire region. This facility could also function as a warehouse or market for national and international buyers.

8.1.3 *Fruit drying*

Dried fruit is fruit from which the majority of the original water content has been removed either naturally, through sun drying, or through the use of specialized dryers or dehydrators. In our case, the dehydration process is based on geothermal waste heat.

Today, dried fruit consumption is widespread. Nearly half of the dried fruits sold are raisins, followed by dates, prunes, figs, apricots, peaches, apples and pears. These fruits are usually dried in heated wind tunnel dryers.

For drying uses, the heat energy from a 50 MWe geothermal power plant (100 MWth) would suffice for a factory which could

— Dry 50 tons of food per hour
— Fill around 25 cargo shipping containers per day.

Needless to say, this facility would likely encourage farmers to enter into previously unknown production of fruit and other kinds of food that can be dried. To some extent, fruit drying would probably remain a seasonal business and not create revenues all months of the year.

All figures above are indicative. For each country and geothermal field, it will have to be evaluated which uses of waste heat could present feasible options in terms of the infrastructure available, people and their needs and capacity in the areas. There might be various options which are not included in this overview and will have to be discussed in more detail in order to be integrated into the conceptual design of the power plant and the entire project.

8.2 *Profitability of waste heat use by an industrial park*

The following graph (Fig. 14) shows an example of how the waste heat from a 50 MWe geothermal power plant could be used with a nearby sustainable industry park. Using clean energy could be one of the first steps towards receiving environmental and sustainable certification for the producers and the products being made in these facilities. This would make the products suitable for the rapidly growing global markets for sustainable and organic products and could increase their economic value even further.

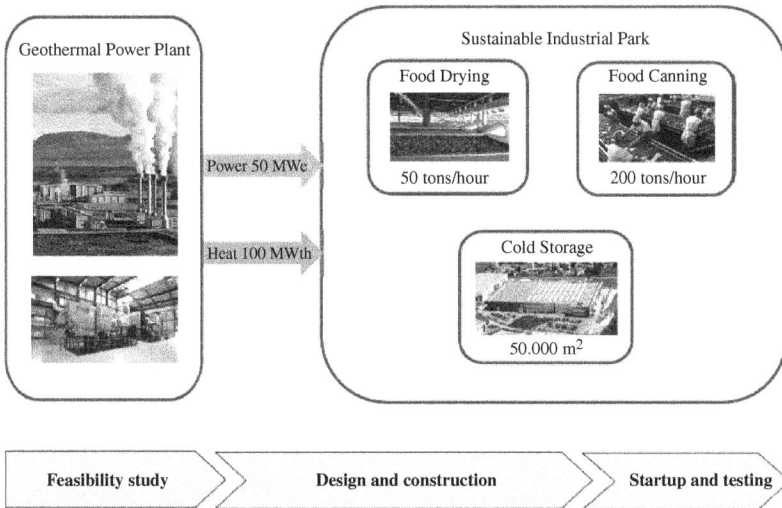

Fig. 14.   Waste heat use by industrial park.[10]

The facilities described above and included in the suggested Industry Park would have the following annual turnover (indicative figures), based on their deployment in the Park:

- 40% Canning: $2 m/day → $120 m/year
- 40% Cold Storage: Inv. $ 200 m → $75 m/year
- 20% Drying: $1/kg → $55 m/year

The total Waste Heat Industry Park: → $250 m/year

Total turnover of power generation at 50 MWe:

$$50,000\,\text{kW} * 8,000\text{ hours} * 8\text{ US\$cents} → \$32\text{ m/year}$$

Comparison on total turnover: Waste Heat: Electricity = 8:1

8.2.1   Conclusion for the waste heat case study

The above example of waste heat use by an industrial park shows that revenue creation from waste heat use could be several times the revenues from power generation. At the same time, the use of waste heat supports regional development through job creation and new agricultural opportunities, e.g. through marketing of new products. Finally, this would imply additional tax revenues to the country, increased value of food products and potential new export products.

8.3   Waste heat use as part of the country's legal framework and PPA

In general, a country's government would have three options for handling waste heat projects from geothermal (as well as other sources, like biomass or industries):

(a) Reduce the tariff (to the IPP) for the generated power and add a payment for waste heat.
(b) Pay the normal tariff for power and pay an additional tariff for waste heat.
(c) Pay the normal tariff for power and get the waste heat free.

As previously discussed, the use of waste heat should improve the overall project benefits for the entire country as well as for the people in the region by the creation of additional jobs and the provision of new opportunities. All a power plant operator will have to do is to make the waste heat from the plant available to waste heat users, e.g. a developer/operator of a canning factory, instead of cooling it down and re-injecting it right away. This requires two interfaces for the waste heat users; one, where the waste heat is received from the power plant, and another one where the fluids are returned to the power plant operator after the heat has been extracted. This process reduces the usage of cooling towers and thereby actually can save operational and maintenance costs for the power plant operator. As mentioned earlier, for security reasons, the cooling towers still have to be installed and be instantly operational. For this extra work and extra capital, including providing the two interfaces, the power operator will have to be compensated well enough to ensure his firm and continued interest in providing his waste heat reliably to other users.

Option (c) above does not provide this additional incentive to the power generator. Options (a) and (b) have to be investigated and negotiated on a project-by-project basis. For example, if the waste heat users are readily available and can guarantee a certain off-take of heat from the power plant, thereby providing guaranteed revenue to the power plant operator, option (a) might be most beneficial for all stakeholders of the project.

In most cases, it will take time to identify industries willing to use the waste heat available, and the power plant operator has in fact little control over when this will happen and how secure the payments from these industries will be. Therefore, option (b) will likely be the most commonly used option. As mentioned earlier, the tariff paid for every kW of waste heat should be just high enough to ensure sufficient incentive to the power plant operator to provide it at all times. The reliability of this service is very important, since all waste heat applications, especially freezing plants and cold storage, will need access to the waste heat on a 24/7 basis.

The tariff paid to the power generator for every kW of waste heat should be lower than from all other sources. As a basic formula, it should include

(i) Cost for providing and maintaining the interface (fixed cost + maintenance),
(ii) Labor cost for keeping the cooling towers in stand-by mode (minimal),
(iii) Nominal return to power generator to insure interest (negotiated).

Savings from O&M of cooling towers could be deducted from this amount, thereby largely offsetting cost factor (ii).

In any case, it might be a good plan to integrate referral prices for waste heat into the PPA of projects and have guidelines in the country's regulatory and legal framework, stating that the use of waste heat should be a part of all geothermal power projects. It should also clarify the responsibilities of all stakeholders and provide an indicative frame for costs.

References
1. United States Geological Survey, accessed on December 22, 2011 at www.c nsm.csulb.edu.
2. Global geothermal capacity reaches 14,369 MW — Top 10 Geothermal Countries, (2018). Think GeoEnergy — Geothermal Energy News. Retrieved January 11, 2019.
3. M. Gehringer and V. Loksha (2012). *Handbook on Planning and Financing Geothermal Power Generation*, ESMAP publication, World Bank, Washington DC.
4. M. H. Dickson and M. Fanelli (2004). *What is geothermal energy?* Pisa, Italy.
5. ISOR (Iceland Geosurvey, 2005). Husavik process diagram produced for Orkuveita Husavikur, Iceland.
6. M. Gehringer (2011). *Cost comparison of geothermal and other technologies, GRC proceedings.* San Diego.
7. M. Gehringer (2011). Version of geothermal risk and cost graph.
8. M. Gehringer (2017). Alternative design of geothermal support mechanisms and risk mitigation funds. *GRC proceedings*, Salt Lake City.
9. X. Wang, *et al.* (2012). Drilling down on geothermal potential: An assessment for central America, LAC Energy Unit/World Bank
10. T. Bjarnason (2014). *Geothermal power plant engineer.* Consent Energy LLC, Washington DC.

Chapter 5

Hydropower

Giovanna Cavazzini,*,‡ Pål-Tore Storli†,§ and Torbjørn K. Nielsen†,¶

*Department of Industrial Engineering, University of Padova, Italy
†Department of Energy and Process Engineering,
Norwegian University of Science and Technology
N-7491 Trondheim, Norway
‡giovanna.cavazzini@unipd.it
§pal-tore.storli@ntnu.no
¶torbjorn.nielsen@ntnu.no

Abstract

Hydropower has a very long tradition in many countries and was first
used to provide working power for grinding corn, sawing timber and other
previously manual tasks. The development of the modern turbine took
a big step forward in the 17th century when Leonard Euler presented
his turbine theory. Later, hydropower became one of the main resources
for electrical energy and it is the most effective method of energy transformation with efficiencies of modern Francis turbines being above 95%.
Even if hydropower already plays a key role in the energy production
contribution from renewable energy sources with an impressive worldwide installed hydropower capacity, there is still a huge potential to be
utilized throughout the world. This chapter will describe hydropower
plants and technologies with an insight on the energy conversion principle and physics. The benefits for the grid as well as the environmental
aspects will also be discussed.

1 Introduction

As the world is looking to replace fossil-based fuels with renewable sources
of energy, there is an increasing focus on the development of new renewables.

Currently, solar PV and wind turbines have shown the greatest penetration. There is, however, one mature renewable source of energy that has been around for decades, even centuries (in fact, millennia!) already. This technology is Hydropower and has evolved from originally being developed for milling grain into highly efficient machines turning solar power used to drive the water cycle of our planet into electricity.

This development accelerated, as for most technologies, during the 1700s and 1800s due to giant leaps in calculus, physics and material science and technology. The famous mathematician, Leonard Euler, developed the equations on how to perform calculations on rotating machinery subjected to a moving fluid. The discovery of electromagnetics and the inventions leading up to the electrical generator made it possible to install machines that extracted energy from moving water, transformed it into electricity that could easily be transported vast distances, and subsequently transformed back to mechanical energy in electrical machines or components doing a specific task.

The historical development of water turbines includes many names. Incremental and more radical improvements have been made by people whose names have not been carried on to describe and classify the technologies. Three names have survived up until today, and they represent the three major turbine types used today: Francis, Pelton and Kaplan. The Francis turbine was the result of refinement of the work by several predecessors, and ultimately carried out by James B. Francis. In 1849, he obtained over 90% efficiency on his latest design, and the Francis turbine became the first modern water turbine. About 60% of all energy produced annually from hydropower worldwide is generated by Francis units. A further development of the Francis turbine by Viktor Kaplan around 1913 resulted in a radical new design suitable for locations with high flows but small water surface elevation differences (low "head"). This new design incorporated a turbine which had adjustable runner blades, enabling a high efficiency over a wide operation range. The name of this technology was, in good tradition, the name of the inventor, Kaplan. On the diametrically opposite end of the scale regarding flow and head (low flow, very high head), work led to a turbine type which was based on a very intuitive operation principle; high velocity water jets impact buckets placed on a large rotating wheel, and forced this to rotate. The work leading to a breakthrough based on this principle in 1879 was by Lester A. Pelton, and the turbine type was designated with his name Pelton.

HYDROPOWER INSTALLED CAPACITY WORLDWIDE

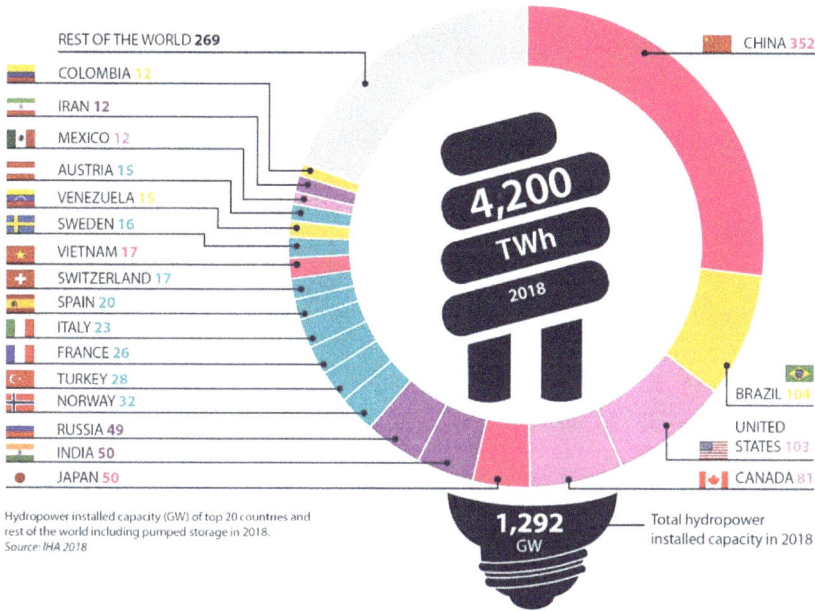

REST OF THE WORLD **269**
COLOMBIA 12
IRAN **12**
MEXICO 12
AUSTRIA 15
VENEZUELA 15
SWEDEN 16
VIETNAM **17**
SWITZERLAND 17
SPAIN 20
ITALY 23
FRANCE 26
TURKEY 28
NORWAY 32
RUSSIA 49
INDIA 50
JAPAN 50

CHINA 352
BRAZIL 104
UNITED STATES 103
CANADA 81

4,200 TWh 2018

1,292 GW

Hydropower installed capacity (GW) of top 20 countries and rest of the world including pumped storage in 2018.
Source: IHA 2018

Total hydropower installed capacity in 2018

Fig. 1. Hydropower installed capacity of top 20 countries and rest of the world, including pumped storage in 2018.[1]

2 Existing Hydropower WorldWide

Today, everything seems to indicate that hydropower and pumped-hydro is experiencing a second or even a third youth. The main drivers of such reemergence are mostly related to concerns on climate change and the depletion of fossil fuels, which have been transposed into international agreements, regional directives and national and local regulations. Because of these concerns and associated regulations, there has been a strong deployment of renewable energy (REN) technologies, which is still ongoing.

In such a context, hydropower is undoubtedly one of the most mature technologies with an annual electricity generation of 4,200 terawatt hours (TWh) in 2018, the highest ever contribution from a renewable energy source, and with a worldwide installed hydropower capacity of 1,292 GW[1] (Fig. 1). In 2018, the other renewables combined only contributed 9.7%, which is about two-thirds of the hydropower contribution to the global electricity production (15.9%) (Fig. 2).

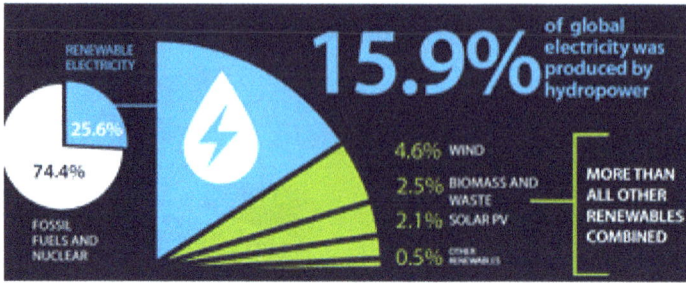

Fig. 2.   Electricity production in 2018.[1]

HYDROPOWER GROWTH THROUGH THE DECADES

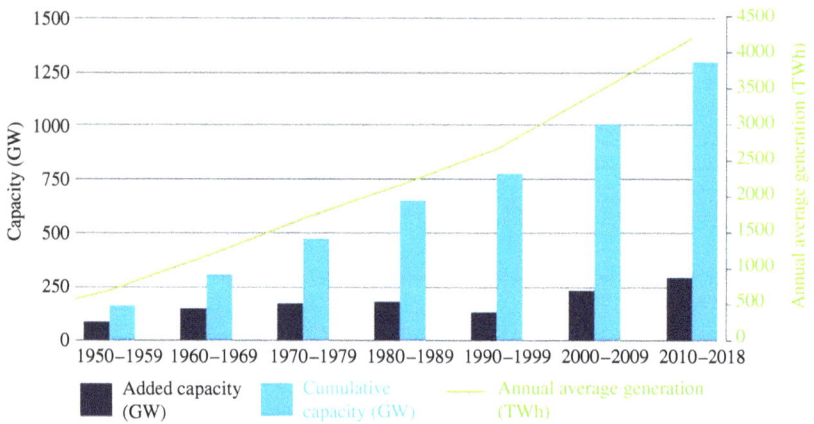

Fig. 3.   Hydropower growth through the decades.[1]

Looking at the hydropower growth through the decades, even if hydropower has always played a primary role, it is interesting to note that its growth faced a contraction in the 1990s, mainly due to environmental concerns (Sect. 6) followed by a second youth in the last 20 years with a doubling of the added capacity (Fig. 3).

Regarding the world distribution of this added capacity: The greatest amount was installed in East Asia and the Pacific Region (9.2 GW in 2018) with the main role played by China with an added capacity of 8.5 GW. This region was followed by South America (4.9 GW), South and Central Asia (4.0 GW), Europe (2.2 GW), Africa (1.0 GW) and North and Central America (0.6 GW)[1] (Table 1).

Table 1.   New Installed Capacity in 2018: Top 20 Countries.

Rank	Country	Capacity added [MW]	Rank	Country	Capacity added [MW]
1	China	8,540	11	Austria	385
2	Brazil	3,866	12	Cambodia	300
3	Pakistan	2,487	13	Laos	254
4	Turkey	1,085	14	Zimbabwe	150
5	Angola	668	15	USA	141
6	Tajikistan	605	16	Iran	140
7	Ecuador	556	17	Congo	121
8	India	535	18	Colombia	111
9	Norway	419	19	Peru	111
10	Canada	401	20	Chile	110

Table 2.   Pumped Hydropower Storage Capacity (GW) of Top 10 Countries and Rest of the World in 2018.

Country	1 Capacity [GW]	2 Country	3 Capacity [GW]
China	30.0	Germany	6.8
Japan	27.6	Spain	6.2
USA	22.9	Austria	5.5
Italy	7.6	India	4.8
France	7.0	South Korea	4.7
		Rest of the world	37.2
		Total Capacity	160.3

In such a context, renewed interest is also growing in large Pumped Storage Hydropower Plants (PSHP), the so-called "water battery", and a huge demand for the rehabilitation of hydropower plants is emerging globally both due to further increases in the corresponding share of renewable electricity production and due to the support, in terms of storage capacity, of a wider exploitation of other renewable energy sources such as wind and solar power.

Presently, PSHP accounts for over 94% of the installed energy storage capacity (160.3 GW in 2018) and over 99% of the energy stored. Even in this case, China retains the largest storage capacity (30 GW) followed by Japan (27.6 GW), the USA (22.9 GW) and the main European Countries[1] (Table 2).

Despite this exploited capacity, it is estimated that a great potential is still available both for hydropower and pumped-storage hydropower plants.

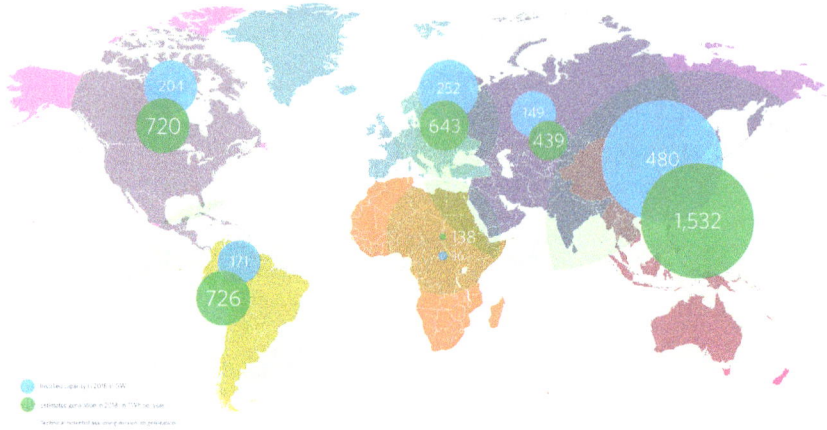

Fig. 4.　Hydropower capacity and generation by region: installed capacity (blue bubbles), estimated generation (green bubbles), technical potential assuming maximum generation (light green bubbles).

In the map reported in Fig. 4, it clearly emerges that the most significant share of this untapped potential is in the Eastern part of the world, where the great availability of water and proper terrain and ground characteristics create the perfect combination for a further increase in hydropower contribution to the world's electric energy generation.

After having improved its power transmission lines, China is now proceeding with the activities related to the exploitation of hydropower: In 2018, the 1,200 MW Shenzhen pumped-storage hydropower station was commissioned, and the 600 MW Qiongzhong station has recently entered into operation. Moreover, China began construction of three pumped storage projects (1,200 MW Fu Kang, 1,800 MW Jurong and 1,200 MW Yongtai), and proceeded with the construction of the 16,000 MW Baihetan project and commissioned several conventional projects, including the 1,900 MW Huangdeng, the 348 MW Sha Ping II, the 920 MW Dahuaqiao and the 420 MW Li Di stations.

Important roles in hydropower growth are also played by North and South America, where the generation of electricity from hydropower is expected to double in the next several years. In particular, Brazil contributed 80 per cent of the region's added hydropower capacity and is the second fastest growing country in the world for hydropower after China. A particular note which has to be made of Africa, having the highest percentage of untapped technical hydropower potential in the world. The total installed capacity in 2018 was equal to 36.3 GW and it is expected to grow by over 40 GW in the next two years. Major projects include the 2,070 MW Lauca hydropower project in Angola (1,336 MW already operational), the 750 MW Kafue Gorge Lower Power Station in Zambia, the 700 MW Zungeru project in Nigeria, as well as the 183.2 MW Isimba project, commissioned in March 2019, and the 600 MW Karuma project in Uganda, soon to be commissioned.

It is clear that the characteristics of the sites in the Americas, in Africa and in the different Eastern Countries are extremely different in terms of availability of water, terrain characteristics and water management strategies. For this reason, there is a large variety in how hydropower technologies are applied. The major differences will be explained in the following section.

3 Hydropower Technologies and Plants

A hydropower plant needs a structure of some kind allowing water to be pulled by gravitational forces from an elevated water body, through a machine called a "hydraulic turbine", to a lower elevation (Fig. 5). The hydraulic turbine extracts energy from this water flow and electricity is produced and distributed via an electrical grid.

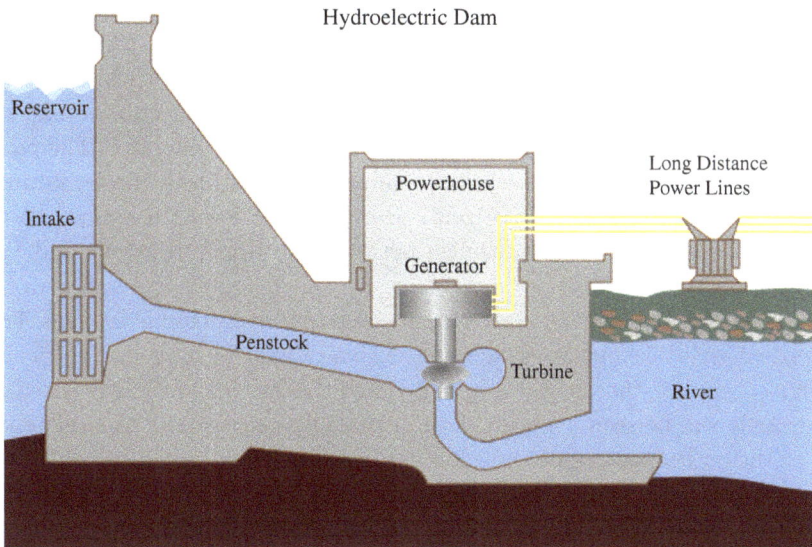

Fig. 5.   Simplified scheme of a hydropower plant.

Fig. 6.   Powerhouse of a hydropower plant. Courtesy of 45 Engineering s.r.l. (https://www.45-eng.com/).

The water stored in an upper reservoir (for example a natural or artificial lake) — called "head water" — is guided by a penstock to a turbine and then released downstream (tail water) into a lower reservoir or into a river. The turbine converts the gravitational energy of the water into mechanical energy and then into electrical energy by means of the generator. The powerhouse, in which the hydro-electric equipment is installed, can be located at the foot of the reservoir (Fig. 6) or on a lower level (in a cave) so as to increase the elevation difference between the upper reservoir and the tail race.

The hydraulic turbines are characterized by two possible layouts. The "vertical axis" layout in which the turbine shaft is mounted vertically (Fig. 7) and the "horizontal axis", in which the turbine shaft is horizontal.

As it will be seen more in detail in the following section, the difference in elevation between upper and lower reservoirs (often called "Head") is a key parameter for the hydropower plant characteristics and it affects the choice of the correct type of hydraulic turbine (Fig. 8).

For the higher heads, the Pelton hydraulic turbine is typically chosen (Fig. 9).

Fig. 7. Example of a vertical axis layout (Francis turbine). Courtesy of 45 Engineering s.r.l. (https://www.45-eng.com/).

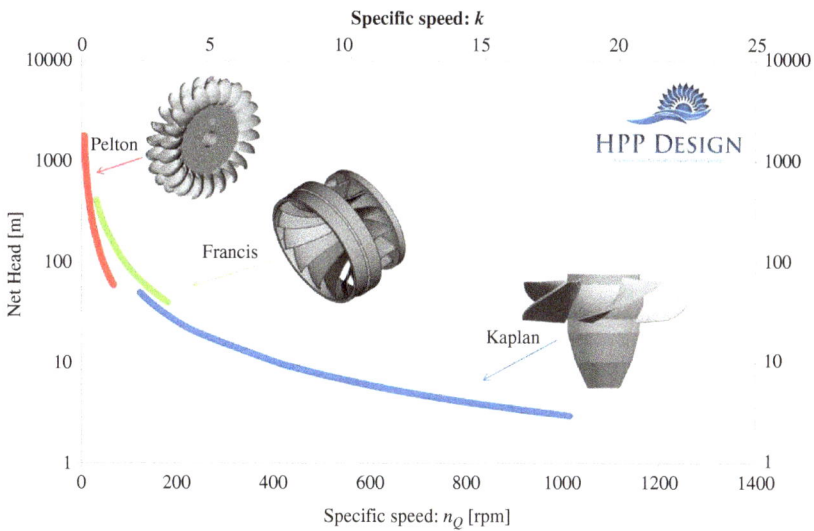

Fig. 8. Influence of the net head on the hydraulic turbine choice (https://hpp-design.com/).

The Pelton turbine is composed of a stator part, characterized by one or more nozzles, and by a runner (also called "wheel") characterized by cup-shaped blades known as "buckets" evenly spaced around the entire wheel's external diameter (Fig. 9). The buckets can be melted together with the

(a)   (b)

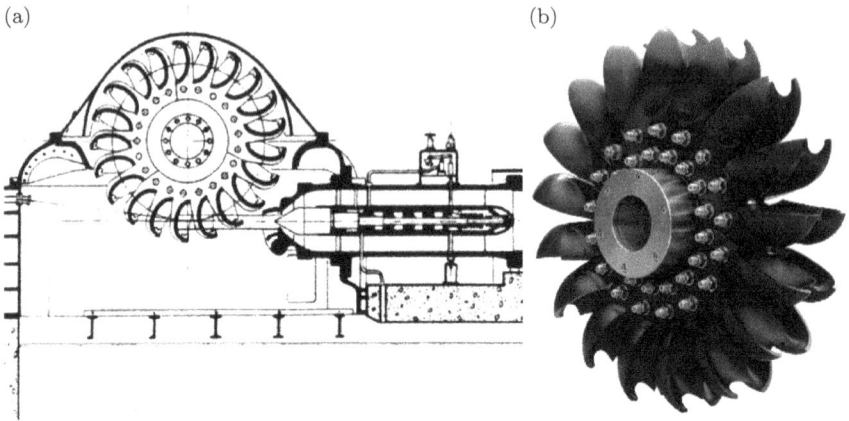

Fig. 9.   (a) One-jet Pelton hydraulic turbine scheme[2] and (b) with a detail of the Pelton runner.

Fig. 10.   Four-jet Pelton turbine. Courtesy of 45 Engineering s.r.l. (https://www.45-eng. com/).

wheel or keyed on it. The water, coming from the reservoir, has transformed its potential gravitational energy into kinetic energy and is pushed through nozzles (one or more than one) into the atmospheric pressure, creating one or more high-speed water jets (Fig. 10).

Each jet hits the buckets at splitters and is divided into two separate streams; then each stream flows along the inner curved surface of the half

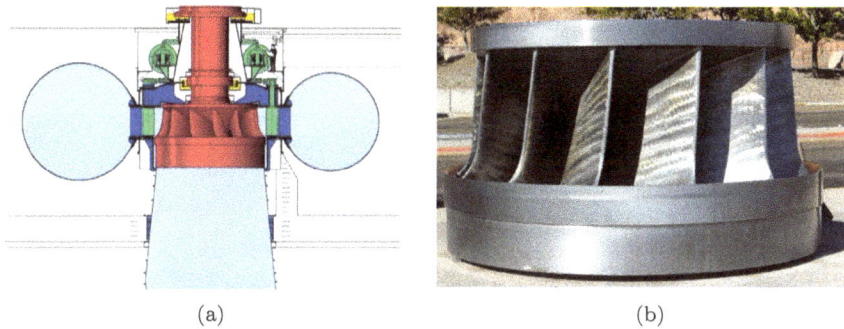

Fig. 11. (a) Francis hydraulic turbine scheme and (b) with a detail of the runner.

bucket and inverts its direction by about 170°. The resulting change in momentum of the water creates an impulse on the blades of the wheel, generating torque and rotation of the turbine.

For medium heads, the Francis turbine is typically chosen. This turbine type currently provides 60% of the hydroelectricity globally. The range of heads for which Francis is typically chosen is wide and this turbine type is commonly further distinguished into high-head and low-head Francis turbines, which visually appear quite different to the untrained eye. They operate however on the same principles for both applications, based on the momentum variation (see Sect. 4).

Fig. 11 shows an example of a medium-head Francis turbine with the detail of a Francis runner. The turbine is composed of a spiral case, a stator cascade of guide vanes surrounding the rotating runner and by a draft tube. The water flow enters in the spiral case, passes through the stator guide vanes and then hits the runner blades with a direction imposed by the guide vane angle, to be lastly discharged in the draft tube. The variation of the momentum is achieved in the runner and it is strictly related to the runner blade leading and trailing edge angles.

For low heads, the turbine types commonly used for low heads are known as Kaplan turbines (Fig. 12). At some locations, the variant known as a bulb turbine is used.

The turbine runner (rotating part of the turbine unit) for both types looks very much like propellers or fans, and is characterized by an axial main flow instead of a centripetal one, characterizing the Francis turbine. Another difference between Francis and Kaplan turbines is that Kaplan turbines generally offer the possibility of modifying the runner blade angle

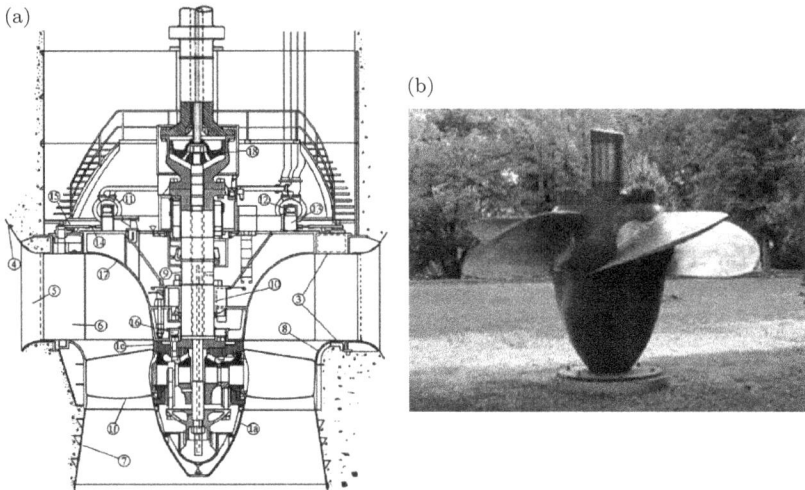

Fig. 12. (a) Kaplan hydraulic turbine scheme and (b) with a detail of the runner.

so as to improve the interaction of the runner blade with the fluid flow, hence achieving higher efficiency values.

As explained, the head plays a key role in the selection of the turbine type, but it is not the only parameter to consider in the design of the hydropower plant. Another key role is played by the flow discharge (volumetric flow), whose product with the head define the available power at a site. The discharge and head (mean value and availability/variability during the year) affect not only the choice and sizing of the hydraulic machine but also the characteristics of the hydropower plant (efficiency, annual energy production, energy and power storage capacity, etc.). For example, some rivers show remarkable temporal variations in discharge with consequent large variations in power. This could affect the power plant performance and the return of investment, since such power plants would not be able to readily obtain a balance between electricity produced and electricity consumed.

So, depending on the site characteristics, different plant solutions should be considered.

3.1 Storage hydropower plants

A storage hydropower plant is an impoundment facility, using a dam to impound water (for example, from a river) and to store it in a reservoir in order to release it later when needed (Fig. 13).

Fig. 13. The Itaipu storage hydropower plant: Overview of the dam built on the Paranà river.

Depending on the storage capacity of the reservoir, it is possible to identify very large reservoirs which can store inflow for months or even years, or seasonal reservoirs, designed for a seasonal storage, to supply water during dry seasons.

Storage hydropower plants are very important for the electrical grid they are producing power for. The water storage provided by the constructed dam is in fact an energy storage system allowing for producing power when needed throughout the day, stabilizing the grid and shifting power production between wet and dry seasons.

The biggest hydropower plants built in the world are storage hydropower plants. Among them the Itaipu and the Three Gorges power plants deserve to be mentioned.

The Itaipu hydropower plant is located on the border between Paraguay and Brazil. Its dam (height: 196 m; length: 7919 m) was built on the Paranà river, flooding the upper reservoir and creating an artificial lake of 29 billion cubic meters of water spread over a surface of about 1400 km^2 (Fig. 13). The plant is characterized by 20 water intakes bringing the water stored in the lake to the generating units (700 MW per unit; 14 GW in total) through penstocks (internal diameter: 10.5 m).

To discharge the water not utilized for generation, the plant is equipped with a spillway, whose maximum discharge capacity is 62.2 thousand m^3/s,

Fig. 14. The three Gorges storage hydropower plant: Overview of the dam on the Yangtze river.

40 times greater than the mean discharge of the Iguaçu Falls, which are a series of magnificent and famous falls very close to the Itaipu power plant at the Brazilian-Argentine border.

The Itaipu hydropower plant is still the world's largest power plant in terms of energy production (world record of 103.1 TWh in 2016), but not in terms of installed power. This record belongs to the Three Gorges hydropower plant (22.5 GW vs. 14 GW).

The Three Gorges hydropower plant, in operation since 2008, is located in the Xilingxia Gorge, one of the three gorges of the Yangtze River, in Hubei province (China) (Fig. 14).

The dam stands 185 m tall and 2,309 m wide, creating an artificial lake of 39.3 billion cubic meters of water. The plant is equipped with a total of 32 main power generators. The first 26 (12 sets on the right bank and 14 sets on the left: 18.3 GW in total) were installed in 2006 and 2008, respectively, and became operational in October 2008. Another six generators were added to the underground power plant of the project and became operational in July 2011.

3.2 Run-of-river hydropower plants

In storage hydropower plants, the dam allows the decoupling of the river flow from the discharge used for power production. On the other hand,

Fig. 15.   Dam of the Isola Serafini hydropower plant on the Po river (Italy).

run-of-river power plants have to be operated in close relation to the instantaneous river flow. This is often because a storage dam can't be established,
typically due to very flat upstream terrain and/or unacceptable consequences when flooding the upstream land.

Large run-of-river power plants are typical of the rivers flowing in
valleys, characterized by an almost constant flow rate during the year.
An example is the Isola Serafini power plant, built on the Po river in the
Po valley (Fig. 15). This power plant (P = 80 MW) is characterized by a
constant flow rate of about 1000 m^3/s and a seasonal head varying between
3.5 and 11 m. The low head value and the high flow rate are the typical
application case of Kaplan turbines and indeed the plant is equipped with
four low-speed Kaplan turbines (external diameter: 7.6 m; rotation speed:
56.8 rpm).

Such power plants still require a dam, whose goal is not to store water
but to divert part of the river flow rate into a channel (Fig. 16). This
"diverting" channel brings it into a loading task, providing a limited storage
capacity, generally lower than 2 hours.

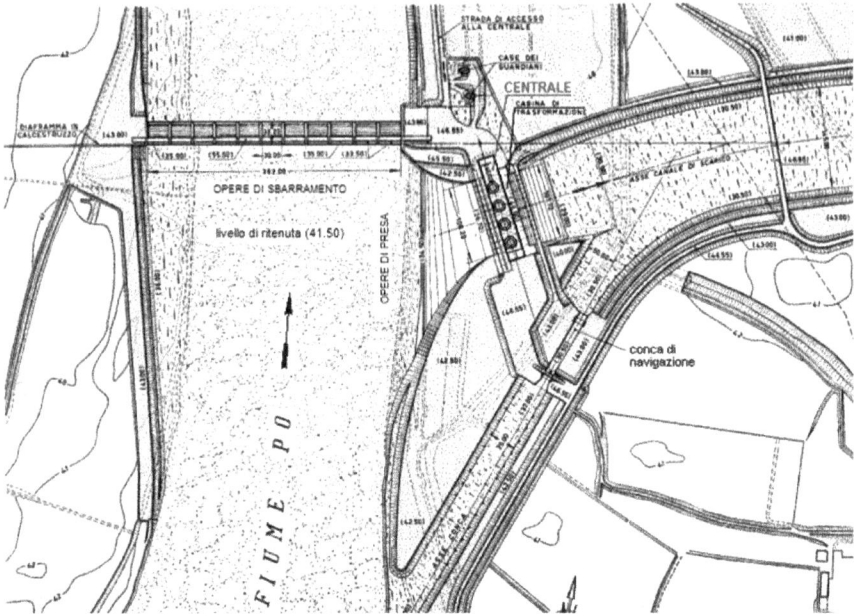

Fig. 16. Scheme of the Isola Serafini hydropower plant on the Po river (Italy).

Because of this small storage capacity, run-of-river power plants have a very limited capability for providing power balancing and other valuable services to the grid. However, they provide energy to the system like wind turbines, whose production is also coupled to the wind availability.

The water is then released downstream from the dam through the channels guiding the tailwater back into the river (Fig. 17).

3.3 Pumped storage hydropower plants

Electrical energy storage becomes an increasingly important issue when the amount of variable Renewable Energy Sources (vRES) increase in a system. The number of ideas and concepts related to providing energy storage has increased exponentially in the last few years due to the shift from electrochemical storage to mechanical storage (Fig. 18).

All these storage solutions present different storage characteristics in terms of timescale, power capacity, energy capacity and response time and all of them can contribute to make the grid safer and more stable by providing different regulation services (Fig. 19).

Fig. 17.   Example of a channel releasing the tailwater into the river.

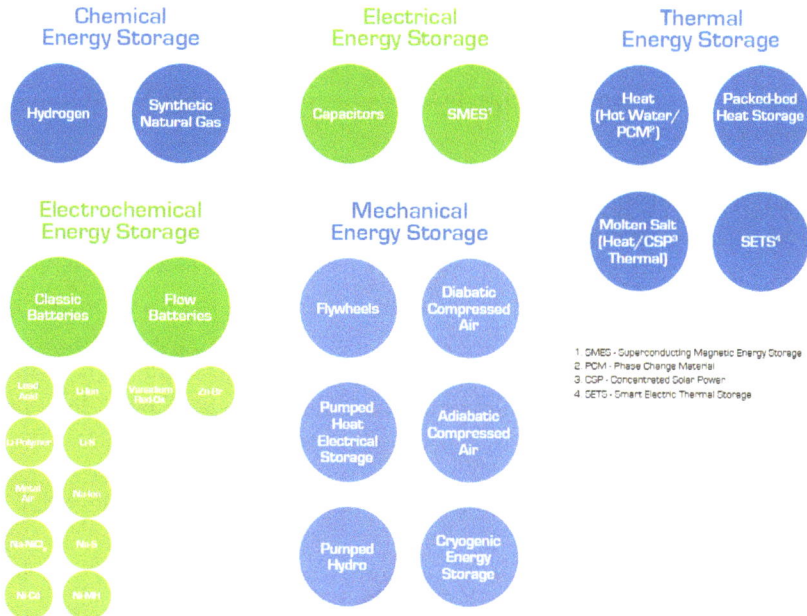

Fig. 18.   Overview of the storage technologies.[3]

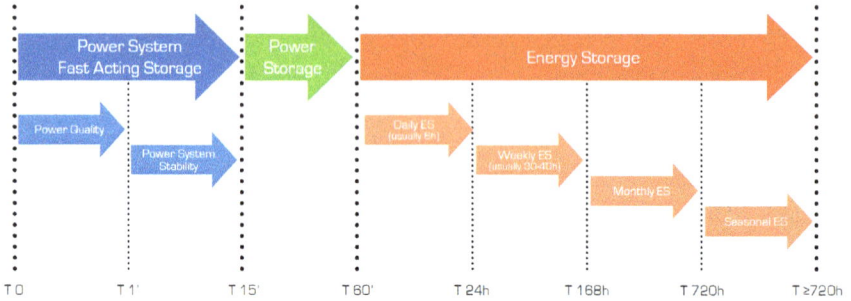

Fig. 19.   Energy storage applications segmented by discharge time.[3]

Among these storage technologies, Pumped Storage Hydropower Plants (PSHP) certainly constitute the most cost-effective technology for boosting power regulation capabilities for plant operators, with competitive costs (300–400 €/kW) and a cycle efficiency in the range of 65–80%.[4,5]

A PSHP converts grid-interconnected electricity to hydraulic potential energy (so-called "charging"), by pumping the water from a lower reservoir to an upper one during the off-peak periods, and then converting it back during the peak periods ("discharging") by exploiting the available hydraulic potential energy between the reservoirs like a conventional hydropower plant (Fig. 20).

These plants require very specific site conditions to be feasible and viable, including proper ground conformation, the difference in elevation between the reservoirs and water availability. For these reasons, the earliest PSHPs were built in the Alpine regions of Switzerland and Austria whose ground conformation together with the presence of hydro resources was suitable for PSHP.

Several machine configurations have been used throughout the history of pumped storage. These configurations differ in the number of hydraulic and electric machines used. In general, they can be classified as:

- Binary set: One pump-turbine and one electrical machine (motor/generator)
- Ternary set: One turbine, one pump and one electrical machine (motor/generator).

Each configuration has its own advantages and disadvantages. In what follows, these configurations will be described.

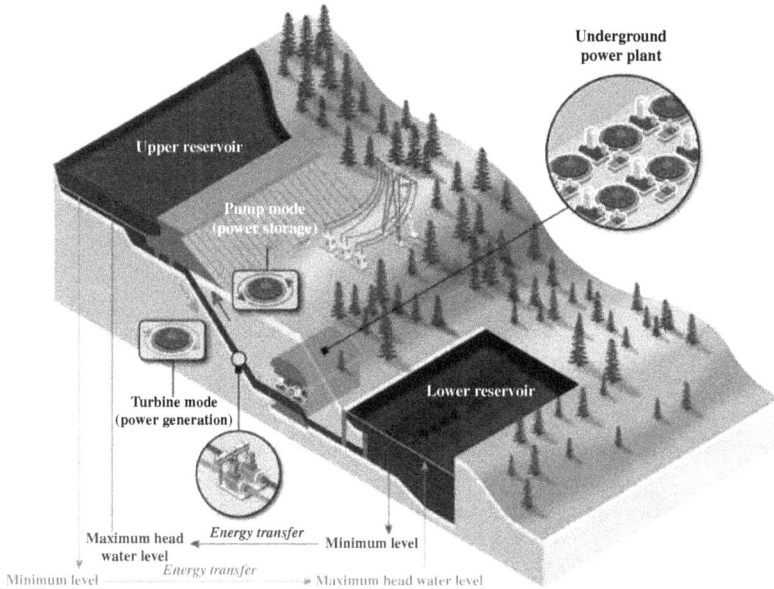

Fig. 20. Scheme of the Pumped Storage Hydropower Plant (PSHP).[6]

3.3.1 *Binary units*

The binary set is undoubtedly the simplest and cheapest configuration and for this reason is the most used scheme in pumped storage hydropower plants.

The configuration is characterized by a reversible pump-turbine coupled to a synchronous electrical machine directly connected to the grid.

The pump-turbine is a hydraulic machine whose design is generally derived from Francis turbines. It is able to rotate in one direction for generating and in the opposite one for pumping (Fig. 21).

The pump-turbine machine can be single-stage or multi-stage depending on the available head (up to 700 m for single-stage; from 700 up to 1200 m for multi-stage configurations) (Fig. 22).

Since this configuration is one of the most used, there are several examples of PHSP equipped with a binary-set. One of the most recent and famous is certainly the Nant De Drance Storage power plant, located in Finhaut, Switzerland. This plant, still under construction, has a head of 311 m and is equipped with 6 single-stage Francis pump-turbines of 900 MW (150 MW each), combined with a variable-speed asynchronous machine (maximum power 175 MVA), rotating at a nominal speed of 429 rpm.

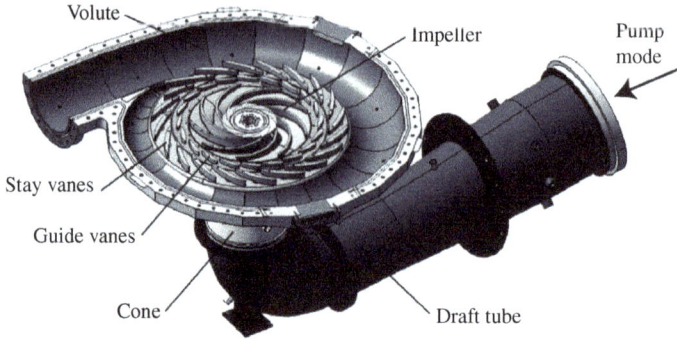

Fig. 21.   Single-stage reversible pump-turbine.

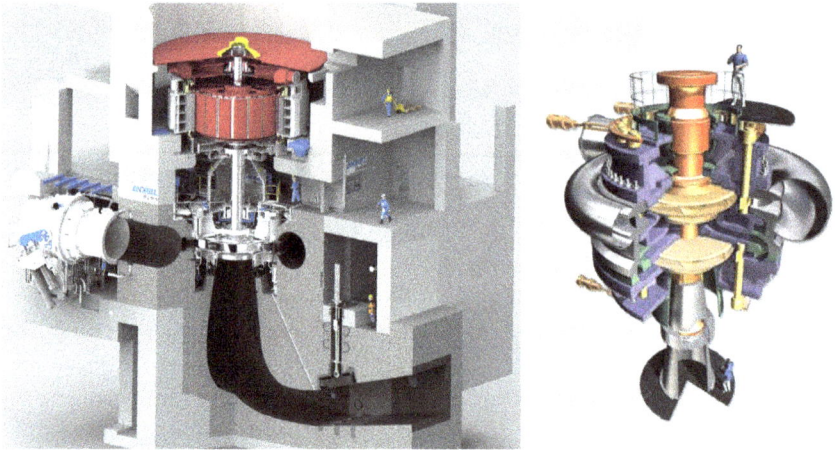

Fig. 22.   Examples of binary set: single-stage (left) vs. multi-stage (right) configuration.

3.3.2   *Ternary units*

In a ternary unit, a turbine, an electrical motor/generator and a pump are coupled together on the same shaft using clutches (Fig. 23). The electrical motor/generator is, usually, a synchronous machine. Different from the binary set in which the pump-turbine design is the result of compromises between the targets of the two operating modes, in a ternary set both turbine and pump designs are optimized with a consequent increase of the plant efficiency.

Nowadays, this configuration is preferred to a binary configuration when the head is high and a single-stage reversible pump-turbine is not

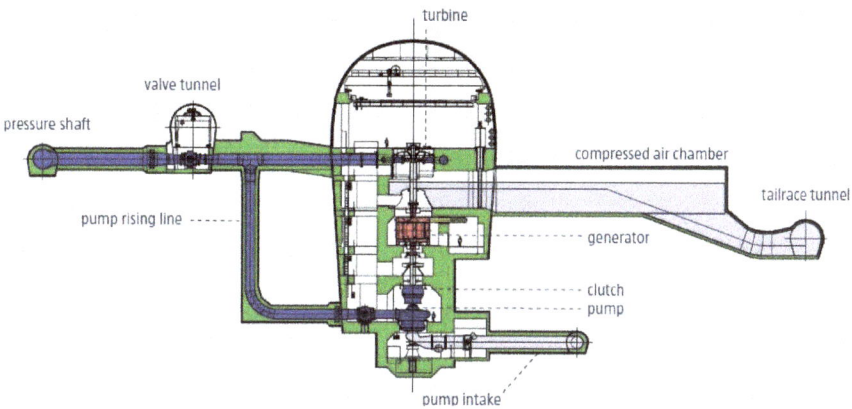

Fig. 23. Kops II power plant scheme (Voralberger Illwerke).[7]

suitable. In this case, a vertical shaft configuration equipped with a Pelton turbine is preferred to a multi-stage pump-turbine. Although both horizontal and vertical shaft configurations could be theoretically used in these cases, vertical ones allow installing the pumps below water level in the lower reservoir and the Pelton turbines above the water level. In order to reduce shaft length, the Pelton turbine can operate inside a compressed air chamber that provides atmospheric conditions in the Pelton runner outlet. An example of this configuration is the Austrian hydropower plant of Kops II (Fig. 23).

Unit operation in generating mode is similar to conventional Pelton unit operation if the pump is decoupled (through the clutch) from the turbine-generator set. It must be taken into account that as these turbines could use deflectors in the water jets, water hammer overpressures can be properly controlled. This is of particular importance in case of an emergency shutdown of the units because deflectors allow eliminating torque in the turbine while the nozzles are closed as slowly as needed.

In order to operate these units as pumps, it is necessary to couple the pump (through the clutch) to the turbine-generator set. Pump start up is carried out with the help of the turbine; once the motor is connected and synchronized to the grid, the nozzles are closed.

In general, times are shorter than in binary units because pump start-up is carried out with the help of the turbine on the same shaft and because changing the shaft rotation direction is not necessary to go from pumping to turbine operation (as it is in binary configuration).

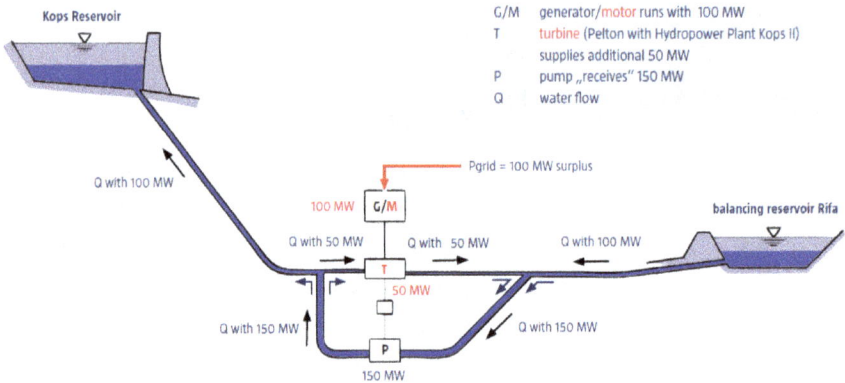

Fig. 24.　Scheme of the hydraulic short-circuit operation in the Kops II PSHP.

Ternary sets also offer more flexible load frequency control solutions in pumping mode in comparison with the binary set by means of the so-called "hydraulic short-circuit operation". With this solution, adopted in the ternary set PSHP of Kops II, the plant simultaneously pumps at the rated power and controls turbine power generation (Fig. 24).

The main disadvantage of this solution is that, for obvious reasons, it implies a larger investment cost, which makes single-stage reversible pump-turbine preferable in case of low- to medium-head and in case of plant refurbishment.

4　Energy Conversion and Physics[a]

Energy comes in many forms, and the ones interesting for hydropower are the ones classified as mechanical energy. For a fluid, this is energy related to pressure, elevation (potential energy) and kinetic energy.

4.1　*Pressure, potential and kinetic energies*

The famous Bernoulli equation describes the mechanical energy in a fluid parcel moving along a streamline in a gravitational field. If no losses are present, the energy is conserved, and the sum of pressure energy, potential

---

[a]The theory is found in Cencel and Cimbala,[8] but most basic fluid mechanics books for undergraduate students have sections describing this.

energy and kinetic energy is constant:

$$\frac{p}{\rho g} + z + \frac{V^2}{2g} = C \tag{1}$$

where p is pressure, ρ is density, g is gravitational acceleration, z is the elevation, V is the velocity and C is a constant value.

This equation is very useful for understanding the interrelation between these energies in a flow, but since actual flows must be considered as an ensemble of fluid parcels subjected to energy losses, the Bernoulli equation must be modified to take this into consideration. Modifications are also needed to consider energies added (pumps), or removed (turbines), from the flow. The modified equation is known as the Energy Equation for flowing fluids, and is

$$w_{pump} - w_{turbine} + \frac{P_1}{\rho g} + z_1 + \frac{\alpha_1 V_1^2}{2g}$$

$$= \frac{P_2}{\rho g} + z_2 + \frac{\alpha_2 V_2^2}{2g} + \sum losses \tag{2}$$

where w_{pump} is the power added to the flow by a pump, and $w_{turbine}$ is the power taken out of the flow by a turbine. The terms α_1 and α_2 are correction factors for the error in using the average velocity V (= volumetric flow rate/cross-section area = Q/A) to represent the average kinetic energy of the flow. This equation is valid between two positions in a fluid moving from point 1 to point 2, in between which there might be a pump or a turbine. If one chooses point 1 to be fixed at the free surface of the water at the upper reservoir and point 2 at the position of a device following the flow and measuring the mechanical energy content (it would be a magnificent device!) of the flow, the device would give the mechanical energy content (in unit meters) of the flow as a function of the position. The line which could have been drawn representing this mechanical energy in a figure representing the layout of the power plant is called the Energy Grade Line. This is seen in Fig. 25.

The Energy Grade Line gives a very visual impression of the losses in the system, because they are represented as the slopes and discontinuities of the line, except for across the turbine element where energy is extracted and not just lost. However, there is a variant of the EGL that is equally useful for engineering purposes, namely the Hydraulic Grade Line. It is quite simply the kinetic energy head subtracted from the EGL. What is left is known as the piezometric pressure (in unit meters), and plotting this

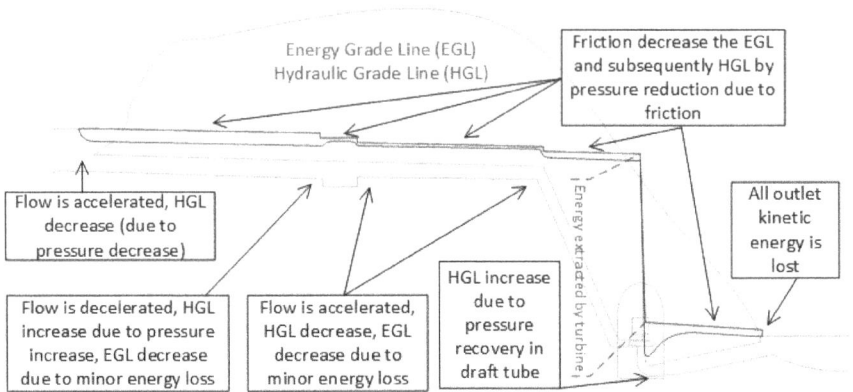

Fig. 25.   Energy Grade Line (red, EGL) and Hydraulic Grade Line (blue, HGL) of a generic hydropower.

in the same figure as the EGL gives new important information. This information is based on the fact that if the HGL at a point or section becomes lower than the elevation of a pipe, tunnel, component or the like, the pressure in that point or section is lower than the atmospheric pressure. If this occurs in a system, there is the danger of air being sucked into the flow causing problems elsewhere in the system. Air into the flow might seem like a small problem, but it might eventually cause oversaturation of oxygen in the water and subsequently diver's disease and deaths occurring in fish downstream a power plant. Air might actually accumulate in parts of the conduit system and if operating conditions change, this air might be released in an uncontrolled and dangerous manner. Furthermore, the air content might affect the pressure wave speed of the system, causing wave propagation frequencies to match with natural frequencies. In this case, undesired phenomena might occur and cause huge trouble for the operation of the power plant. Finally, a bad seal or broken component in parts of a system where the internal pressure is lower than the atmospheric pressure will not be revealed during operation since air is pulled in. But if the process involving the flow is stopped for some reason, the pressure will increase and then a leak will occur. Not all liquids are harmless, so special attention should be given to low-pressure regions in case of dangerous liquids.

As mentioned, the losses in the system will lead the EGL (and also the HGL) to decline. But to be able to draw these lines, the losses must be quantifiable. How to do this will be explained in the next section.

4.1.1 *Energy losses*

"Energy losses" is a contradictory term. Modern science is based on the fundamental hypothesis that energy cannot be produced or destroyed, i.e. *energy is conserved*, but can shift between different forms. Following this argument, it might be tempting to state that "there are no energy losses, only energy transformations!". And this is of course true. But when the "interesting" energy terms are the ones expressing mechanical energy, as is the case for hydropower applications of fluid mechanics, there will be portions of this energy that is "lost" to other energy forms that are not interesting. So, the term "energy losses" might not be that improper after all, even if "mechanical energy losses" is a vastly more precise term to be used in this context. Still the imprecise term "energy losses" is embedded in the terminology of fluid mechanics, and we will not debate its use any further. We will, however, describe how to quantify these (mechanical) energy losses in the following.

4.1.1.1 Friction (major) losses

The friction losses are losses occurring in the pipes, tunnels and conduits of the power plants. These losses can be found from empirical relations, and the friction head loss, usually denoted h_f[m], can be found using the Colebrook–White equation:

$$h_f = f \frac{L}{D_h} \frac{V^2}{2g} \tag{3}$$

where L is the length of the water way, D_h is the hydraulic diameter of the water way (the same as the internal diameter in case of a circular pipe) and f is the so-called Darcy–Weissbach friction factor. The key parameter for finding this factor is the roughness of the water way and the Reynolds number, Re:

$$Re = \frac{\rho V D}{\mu} \tag{4}$$

where μ is the dynamic viscosity of the water. The Reynolds number is the ratio of inertial forces to viscous forces, and is paramount in order to identify the flow regime. For high Reynolds number flow, the viscous forces, implicit frictional forces, are much lower than the inertial forces, but still cannot be neglected for tunnels and conduits of practical lengths. The relation between f and Re is not mathematically simple. It is different for different flow regimes, typically classified as laminar ($Re < 2300$) or turbulent ($Re > 4000$). Flows in hydropower plants are always turbulent because the diameter, velocity and density are high and the dynamic viscosity is low, giving $Re > 10^6$ for hydropower applications. The equation which most accurately describes the friction factor for turbulent flow is the Darcy–Weissbach equation. This is however an equation which implicitly describes f, which requires an iterative scheme in order to solve for f. To overcome this tedious work, the famous Moody chart was developed in 1944, in which the friction factor is plotted against Reynolds number with the relative roughness as a parameter. Still, not all problems are solved because in cases where the flow must be determined, the Reynolds number is unknown, and the chart becomes useless. At the expense of a 2% deviation from the original Darcy–Weissbach equation, the Haaland equation offers an explicit formulation of f. Knowing that the Darcy–Weissbach equation deviates as much as 10% from experimental results, this expense is easily accepted. The Haaland equation is

$$\frac{1}{\sqrt{f}} = -1.8 \log\left(\left(\frac{\varepsilon/D}{3.7}\right)^{1.11} + \frac{6.9}{Re} \right) \qquad (5)$$

where ε is the roughness [m]. In the original Darcy–Weissbach equation, the last term inside the log operator contains the friction factor and makes the equation implicit with respect to the friction factor, making an iterative procedure necessary. This is however the case with many other models of the friction factor as well, and a direct consequence of the strong relation of the velocity on the friction, and the feedback from the friction on the velocity.

4.1.1.2 Singular (minor) losses

Singular losses are related to the energy loss in specific parts of a power plant, i.e. where components such as valves or gates, junctions and cross-sectional changes are located. The singular losses are found using a similar formulation to the frictional losses, although with the term fL/D substituted with a coefficient found from tables, charts or supplied by the manufacturer of the component. The singular loss, usually denoted h_s, is defined as

$$h_s = k_c \frac{V^2}{2g} \qquad (6)$$

The coefficient k_c is dependent on geometry and flow direction. Some coefficients can be seen in Fig. 26. For components where there is a change in

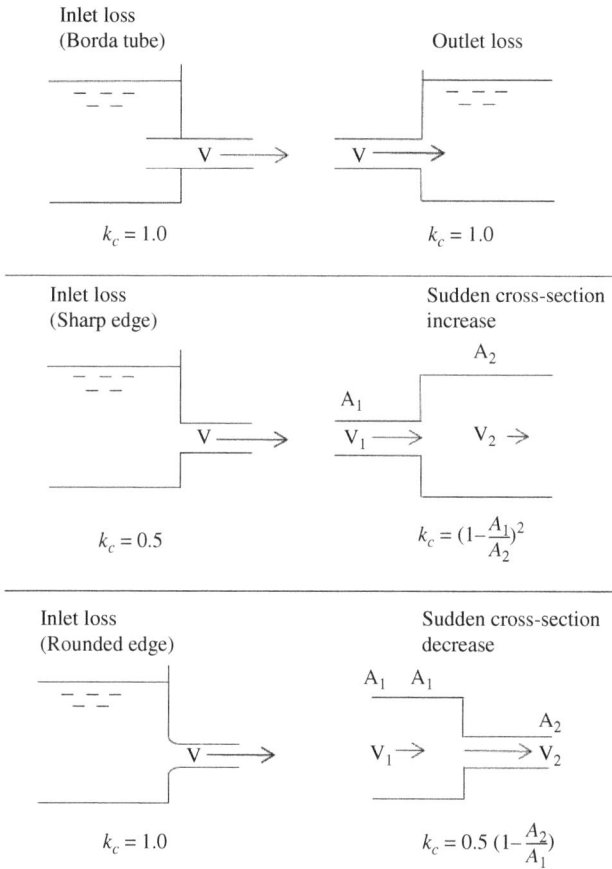

Fig. 26.   Loss coefficient for some pipe interconnections.

velocity for the flow through the component, the highest velocity is typically used in Eq. (6).

Extracting all losses from the available energy head H_g gives the net energy head, H_e, which is available for the turbine to extract as mechanical energy:

$$H_e = H_g - \sum losses \tag{7}$$

In a design phase, one can choose to reduce velocities by increasing the cross-section areas of components, conduits and tunnels, thus reducing the losses. This is however very expensive, as the dimensions of the components increase, more rock must be excavated and deposited, and so on. A techno-economic analysis must be performed in order to find the solution that gives the best overall design and layout of the system. Such analysis is very complex and is beyond the scope of this text. However, large hydraulic power plants typically end up having a loss in the system (excluding the turbine) of about 5–10%. This means that the net head which is available for the turbine is 90–95% of the gross head. This net head, along with the flow used to calculate the losses, will represent the power that is extracted by the turbine unit from the flow. The energy conversion in this process is subjected to fundamental losses and will be described in the next section.

4.2   Energy conversion in turbines

The energy conversion involved in extracting mechanical energy from a flowing fluid is subject to engineering decisions, but as we shall see later there are different reasons for choosing one over the other. One thing which might be useful to remember is that the only way forces can be applied to a structure from a fluid is by either pressure forces (perpendicular to the surface) and/or viscous forces (parallel to the surface). The viscous forces generally do not contribute to rotating the turbine runner about the rotational axis and for conventional turbines are just an energy loss. The curious exemption is the Tesla turbine, which only uses viscous forces, but will not be described in this document.

The available hydraulic power, P_h, in a flowing fluid through a cross-section of a pipe is described as the flow rate Q going through the cross-section multiplied by the pressure P in the cross-section. The power removed from the flow between two cross-sections is the difference in the power flowing through each cross-section. For a turbine, the flow out of it is the same as the flow into it, so the hydraulic power P_h removed from the flow as it passes through the turbine is the flow Q going through the turbine multiplied by the difference in pressure $\Delta P = \rho g H_e$ on each side of the turbine.

$$P_h = Q\Delta P = \rho g Q H_e \qquad (8)$$

Power is extracted from the flow, and as energy (and power) is conserved, it must go somewhere. This is the entire purpose of the turbine: transforming power in the hydraulic domain to power in the mechanical domain. The flow through the turbine imposes a torque on the runner, and the torque T multiplied by the angular velocity ω of the runner/axle assembly is the mechanical power used to drive the rotation. The power transformation from the hydraulic to the mechanical domain is not without losses, so an efficiency η_t needs to be included describing the conservation of power.

$$\omega T = P_h \eta_t = \rho g Q H_e \eta_t \tag{9}$$

If no torque was acting on the runner assembly to balance this torque, the runner assembly would accelerate. On the generator side of the axle, the induced electrical current is generating a magnetic field which imposes a torque similar in magnitude to the one from the water. This is the next step in the energy conversion chain, as the induced current, I, requires a voltage, U, to drive it, and the power in the electric current is voltage multiplied by current. This conversion is not without losses either, so a new efficiency η_g must be included to describe the conversion.

$$UI = \eta_g \omega T \tag{10}$$

The conversion and flow of power can be seen indicated in Fig. 27, which also indicates the friction and singular losses, as they are responsible for H_e being slightly smaller than H_g.

Still, we have not provided any detail on the mechanisms that lead to the power P_h being successfully transformed into mechanical power. To be able to do this, we have to look into the physics of the flow.

For power to be extracted from a rotational movement, torques must be involved. If the rotation is to occur at a constant angular speed, there must be one or more torques trying to accelerate the rotation that combined

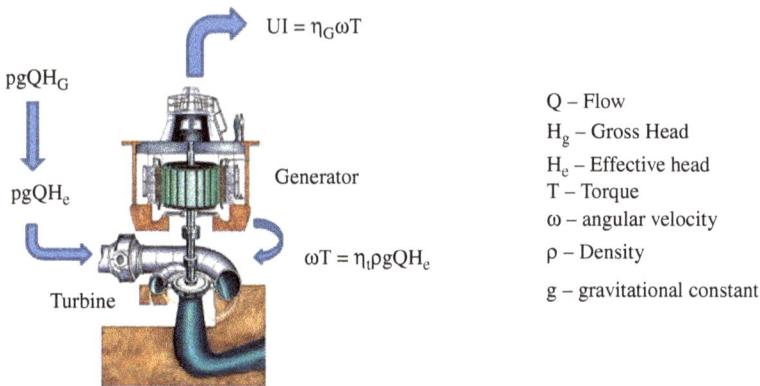

Fig. 27. Energy conversion through a turbine unit, from Potential energy to electrical energy.

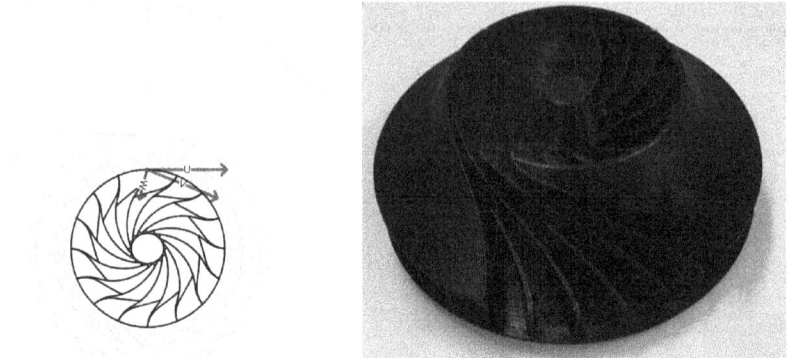

Fig. 28. Axial cut-away schematic of a turbine (Francis); runner in black (left); a small 3D printed model showing how the geometry looks like, upside down (right).

has the same magnitude as the torques trying to decelerate the rotation. The torque trying to accelerate a turbine runner is the one provided by the flowing water into the inlet of the turbine which has an offset with the rotational axis of the unit. This can be seen in Fig. 28, where the flow enters the so-called spiral casing of a turbine. This off-set initiates a swirling motion, and for the flow through the first parts of the unit (spiral casing, stay vanes) the angular momentum is conserved as no energy is extracted from the flow. Through the next part, the guide vanes, the angular momentum is increased because there is an acceleration of the flow in the tangential direction. This acceleration is because the guide vanes are forcing the flow in a different direction than the one it naturally wants to go in, and this acceleration happens at the expense of the pressure dropping. Eventually the water enters the rotating part, i.e. the runner. Here, the water has an angular momentum, and the purpose of the runner is to extract all this angular momentum and in the optimum way to transform this to a torque acting from the water on the runner.

4.2.1 The Euler (pump and) turbine equation

Momentum conservation laws are used in fluid mechanics to describe the link between what drives a fluid flow and its acceleration. For linear motion, there is the "linear momentum equation" explaining the relation between forces and linear momentum of the flow. For rotational motion, we have the "angular momentum equation" describing the relation between torques and angular momentum of the flow. For a turbine which is rotating, the angular momentum equation is the most helpful one. This is a control volume (CV) analysis, where the effect from external torques acting on the water inside the volume must balance the change of angular momentum of the flow, passing through inlets and outlets over Control Surfaces (CS):

$$\sum \vec{T} = \frac{d}{dt} \int_{CV} \rho(\vec{r} \times \vec{V}) d\forall + \int_{CS} \rho(\vec{r} \times \vec{V})(\vec{V_r} \cdot \vec{n}) dA \qquad (11)$$

where an arrow above a symbol means that it is a vector, $\vec{r}$ is the radial position, $\vec{V_r}$ means the relative velocity, in this case between the velocity and the control surface where mass can cross. For simplicity, we can assume that the only torque acting on the unit is the torque acting from the generator, T. If we further assume steady state, uniform conditions on one fixed inlet and one fixed outlet with subscripts 1 and 2, respectively, we get

$$T = \rho Q(r_1 V_{u1} - r_2 V_{u2}) \qquad (12)$$

where V_{u1} and V_{u2} are the vector components of V_1 and V_2 is the direction of U. The right-hand side is the torque extracted from the flow, and the left-hand side is the generator torque. For a wind turbine, the air approaches with zero swirl, meaning that it must leave the outlet of the turbine with a swirl, if a generator is to extract energy from the flow. In turbines, the control of the water enables the water entering the runner to have swirl, and then leave with no swirl. Designing the unit for no outlet swirl is one of the key design philosophies for turbines (see Figs. 29 and 30).

If no swirl is assumed at the outlet, the latter term on the right-hand side is zero. It might be easy to think that you can get as high a torque as you want just by designing a machine that has a high velocity at a high inlet radius, thus getting a high value for the product $r_1 V_{u1}$. In theory you could, but because the power can never be higher than the hydraulic power, we

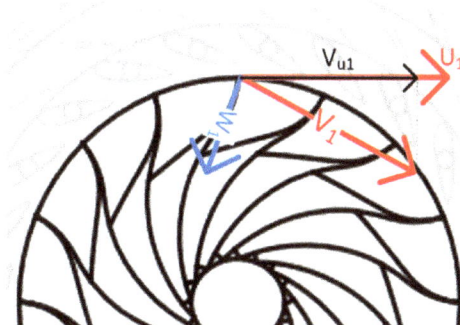

Fig. 29.   Inlet velocity diagram.

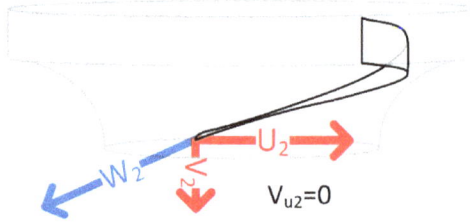

Fig. 30.   Outlet velocity diagram.

need to multiply the equation describing the torque by the angular velocity. We then get

$$T\omega = \rho Q(u_1 V_{u1} - u_2 V_{u2}) \tag{13}$$

This is a very useful equation in the design phase of a hydraulic runner, as the last term containing V_{u2} is designed to be zero at the design conditions. Dividing this by the hydraulic power, we get the definition of the efficiency of the energy transformation.

$$\eta = \frac{T\omega}{\rho g Q H_e} = \frac{u_1 V_{u1} - u_2 V_{u2}}{g H_e} \tag{14}$$

This first part of this equation is very convenient for measuring the efficiency over a wide range of conditions, as the measured properties will deteriorate because of the presence of losses. The last part is not very useful for this purpose, as detailed values of the velocity component are difficult to measure. However, it can be used for developing mathematical analytical models of the efficiency by including models of different fundamental losses occurring in a turbine. These losses are different for different turbine types. The fundamental losses themselves will not be given attention in this text, but the effect they have will be indicated by describing the efficiency curves for the different types. The main three turbine types in use for power generation globally, Pelton, Francis and Kaplan, will be presented, starting with the Pelton turbine.

4.2.2   Pelton turbine

The Pelton turbine is a so-called impulse turbine, meaning that all the available energy is converted to kinetic energy before the water enters the turbine runner, and this happens in a nozzle producing a high velocity water jet. When the water impinges on the runner, the runner geometry

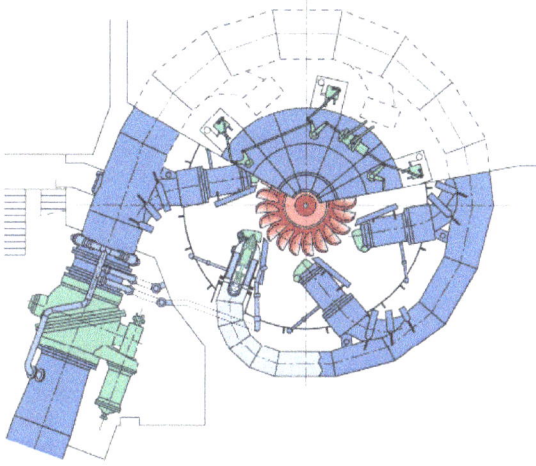

Fig. 31. Cut-away overview of a Pelton turbine. The runner is in center with the so-called buckets seen distributed around the perimeter.

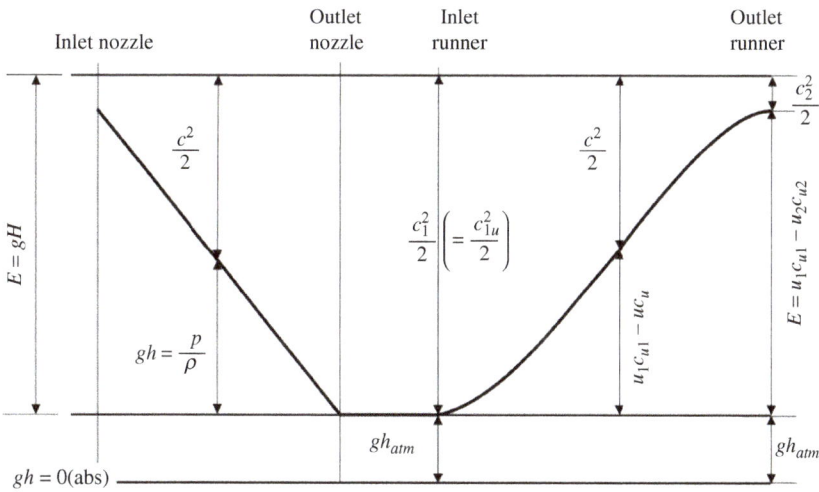

Fig. 32. Energy conversion diagram for a Pelton turbine.

makes the jet almost invert direction (about 170°), leading to a very high water pressure on the runner, responsible for the torque needed to turn it around (see Fig. 31).

The energy conversion diagram for a Pelton turbine can be seen in Fig. 32. The way to understand the diagram is to imagine that the energy terms are tracked as fluid going through a turbine and may be followed in the diagram going from left to right. The black line represents the shift from kinetic energy to other forms of energy. The height from the black line to the uppermost line represents the kinetic energy of the flow at the different positions. The height below the black line is pressure energy and extracted energy. At the inlet of the nozzle, the specific energy is $E = gH$, and most of this is pressure, as the velocity is quite low. Through the nozzle the flow is accelerated until the pressure at the outlet is identical to the atmospheric pressure. This is seen by the increase in height of the line representing the kinetic energy, and the pressure drops to the atmospheric pressure, $P_{atm} = \rho g h_{atm}$. From the outlet of the nozzle to the inlet of the runner nothing really happens, but as water hits the buckets, the extraction of energy reduces the velocity, and through the bucket this continues until there is very little velocity left, just enough to make the water leave the bucket without hitting the backside of the next bucket that follows right behind.

Pelton turbines are used at high heads and have low discharge. The reason is that a Francis turbine designed for the same conditions would have too low an efficiency due to the channels inside the runner becoming long and giving high friction losses. Other difficulties are also present for high head Francis units, so this adds to favoring the Pelton turbine type of unit. At locations with high amounts of sediments in the water, Pelton units are also often preferred due to the easy access to the runner being available for inspection at atmospheric conditions.

The peak efficiency of Pelton runners is the lowest of the three types, but the efficiency is not highly affected by change in discharge, so the efficiency curve is very flat for a wide range of operations.

4.2.3   *Francis turbine*

The Francis turbine is a so-called reaction turbine, meaning that the water entering the runner has both kinetic and pressure energy. So, through the runner, the pressure decreases as well as the kinetic energy.

Francis turbines are made for low, medium and high heads. The runner in Fig. 33 is a typical medium head turbine. The difference between low and higher head runners is that the inlet diameter increases with higher heads.

The energy conversion diagram for a Francis turbine is seen in Fig. 34.

At the inlet of the unit, the energy E is the sum of the pressure energy and kinetic energy. The velocity is uniform and parallel to the axis of the

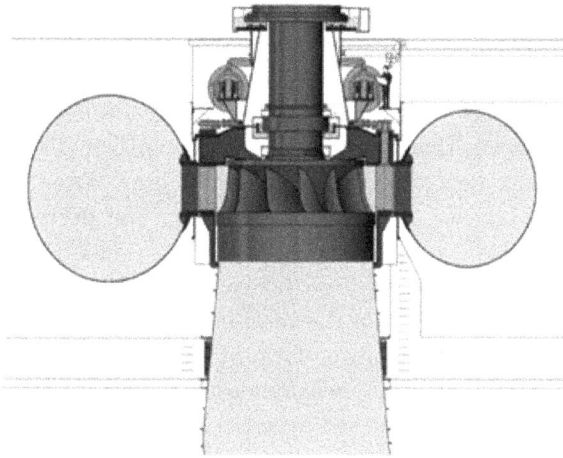

Fig. 33.   Radial cut-away view of a Francis turbine.

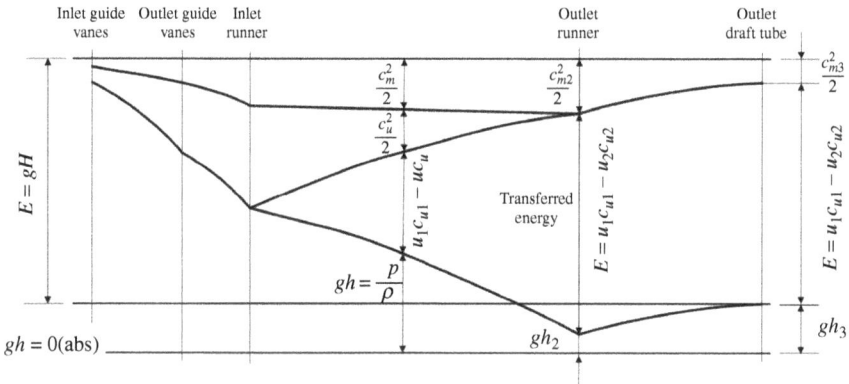

Fig. 34.   Energy conversion for a Francis turbine.

pipe, which means that the kinetic energy is well represented by the flow velocity, $c = Q/A$. As the water moves into the spiral casing, the flow is forced to go into a swirling motion because of the shape of the spiral case. It must also move towards the center of the unit, as the outlet is at a smaller radius than the inlet. This gives a velocity vector that has components both in the peripheral (c_u) and meridional directions (c_m, meridional direction is the direction normal to the local cross-sectional area, meaning it will go from pointing in a radial direction at the inlet to pointing in an axial direction at the outlet). So, when the flow enters the guide vanes, there is kinetic energy linked to the c_u component as well as the c_m component. Through the guide vanes, the c_m component increases due to the flow accelerating because the cross-sectional area is decreasing, and c_u increases because the guide vanes force the water into a more swirling motion. The increase in these velocities means that the kinetic energy increases. This energy comes from an equivalent decrease of the pressure energy. After the guide vane there is still an increase in the kinetic energy, and decrease in pressure energy, before the water enters the runner. This occurs due to the physics involved in a vortex motion. Subsequently, when the flow enters the runner inlet, approximately half of the total energy is in the form of pressure energy, and the other half is in the form of kinetic energy.

The details about what happens inside the runner are corporate secrets, but the overall physics is described by a deceleration of the swirl component along with a pressure reduction.

At the outlet there is the possibility of having rotational velocity components depending on how well the match is between flow, rotational speed and outlet angles of the runner blades. At the design conditions, which commonly are the conditions with the highest efficiency (Best Efficiency Point, BEP), the rotational component of the flow is close to zero, and the peak efficiency of a Francis unit is very high, the highest of the three main types.

4.2.4   Kaplan turbines

The Kaplan turbine is also a reaction turbine. It differs from the Francis turbine with the runner looking very much like a huge fan or a propeller. The same theory that is used for wings is used to design the runner blades, along with cascade theory as they influence each other (Figs. 35 and 36).

The Kaplan turbine is special because it has runner blades that can have a changed angle during operation, making it possible to match the angles of the runner blades to the changing velocity vectors due to operational changes on the guide vanes. The continuous regulation of the runner blades when the entire unit is regulated gives a lot of wear on the runner blade regulation system. The bearings used need to have a lubrication system, and oil leaks have been an issue at some units, the oil going directly out into the environment.

For really high flow units, another version of the Kaplan turbine is used, called a Bulb (or Kaplan Bulb) turbine. It has the main axle oriented horizontally, which makes a draft tube bend no longer necessary and improves

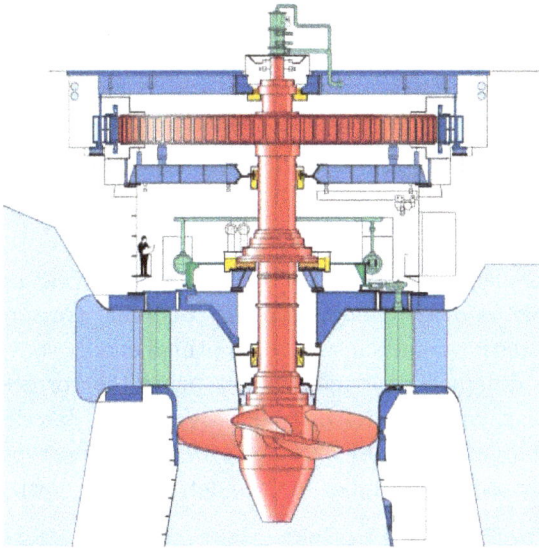

Fig. 35. Radial cut-away view of a Kaplan turbine.

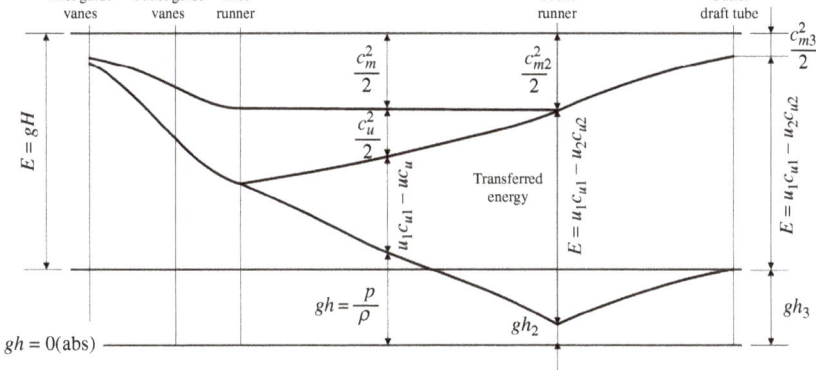

Fig. 36. Energy conversion in Kaplan Turbines.

the draft tube efficiency. These units are so large that it is possible to install the generator directly in front of the runner, cased-in with the water passing around it. This structure looks like a bulb and gives these units their name. The spin of the water previously provided by the off-centered inlet pipe compared to the main axle must still be provided, and large guide vanes in front of the runner provide this.

With fixed runner blades, a Kaplan has a very pointy efficiency curve. But adding runner blades that can be rotated and a cam system that matches runner blade angles to the guide vane angle, the Kaplan turbine has a very flat efficiency curve. The peak efficiency is not as high as the Francis turbine.

4.2.5　*Classification of turbine types*

The large difference in the combination of flow and head means that the energy available is distributed differently between pressure and kinetic energy. This variation results in different optimal energy conversions, which again leads to different types of turbines being the correct ones for the different combinations. Several numbers are used to describe this energy conversion and are used as indicators as to which turbine type is the appropriate one to choose. The number most widely used is known as the specific speed, denoted N_s, defined as

$$N_s = n \frac{\sqrt{Q}}{(gH_e)^{3/4}} \tag{15}$$

where n is the rotational speed of the runner, Q is the flow and H_e is the net head, as described previously. How to find/decide these three quantities, which eventually make computation of the specific speed possible, will be described in the following, starting with the rotational speed n. The rotational speed must be a synchronous speed, meaning it is a fixed speed specific to the generator, and is used in order to supply the electrical grid frequency of the location. The relation is

$$n = \frac{f_g \cdot 60}{p} \tag{16}$$

where f_g is the nominal frequency of the grid at the power plant's location, p is the number of pole pairs (an integer) in the generator. The number of pole pairs can be chosen quite freely, and as a general rule it is cost beneficial to have as low as possible number of pole pairs, giving a bias towards higher rotational speeds which make generators more compact and less expensive. But at high rotational speeds, cavitation (see Sect. 4.3) is more prone to occur, requiring expensive submergence of the turbine unit. A crossing point between these reducing/increasing cost curves will be an optimal configuration. As an example, for a Francis turbine power plant this will typically be the case when the outermost peripheral speed of the turbine outlet, u_2, is between 38 m/s and 42 m/s.

The design value for the discharge Q (flow) can be difficult to decide on. The annual total inflow to the power plant is determined from weather statistics and is not something that can be chosen. Having a large upper reservoir, it is possible to decide the power plant design discharge, as water can be stored and used for production more freely. Some operational strategy or licensing regulations might also change how the power plant design discharge is determined. Once this has been determined, choosing the number and size of units will determine the design discharge for these units. Finding the number and size of the units is not trivial, as the large variety of power plant schemes makes it necessary to have a tailor-made approach. A large dam at a low head site makes it possible to have many units inside the dam construction, and this will give operational flexibility and reduce the cost of each unit due to mass manufacturing benefits. However, at some point the flexibility of additional units will not justify more units being added, so there will be a crossing point between cost and benefit.

A storage power plant will typically end up with a small number of units because of the added flexibility of the storage reservoir. The smallest number of units is one, but a single turbine unit is subject to the cost of a low efficiency, outside the best efficiency operation conditions, as well as the dangers of losing water if for some reason the unit malfunctions or if inspections must be made in the season where production has been assumed to be at full speed. This points to the difficulties of choosing the number and size of units for the specific site conditions. As an example, many Norwegian storage power plants have four identical Francis turbines installed, as this has proved to give the best overall solution. The intended purpose of the power plant will also influence the Q used for the design computation. A storage power plant can be designed so that the total annual inflow is used for power production for only portions of the year. This will give a high Q, but a low duration.

A run-of-river power plant has a very limited storage and must produce power whatever the inflow is to the upper reservoir. This will give a low Q, but a high duration. Such power plants have often several units of very different sizes to be able to obtain a high efficiency for a large variation of flow. No general guidelines can be given in order to find an estimate for the flow, as all projects are unique. Fortunately, for the turbine engineer this decision is out of their hands. They will be presented with an initial number of units and their design flow Q, and will have limited possibilities to alter this.

The net head H_e is in practice defined by the site conditions. A rule of thumb is that the techno-economic analysis for a storage power plant will give 5–10% losses in the hydraulic conduit system. So, the net head will be 90–95% of the gross head.

Now that the rotational speed, flow and net head have been determined, the specific speed can be calculated according to Eq. (16). As mentioned, the energy conversions are different with different combinations of head and flow. Since energy conversion characteristics are built into the specific speed, this number will also be applicable as a characteristic number for the different machines used to optimize the utilization of the available energy.

In Fig. 37, a typical distribution of the most common turbine types can be seen.

4.3 *Cavitation*

Cavitation is a phenomenon that has potentially very negative effects for hydraulic machinery. The phenomenon occurs when the water pressure

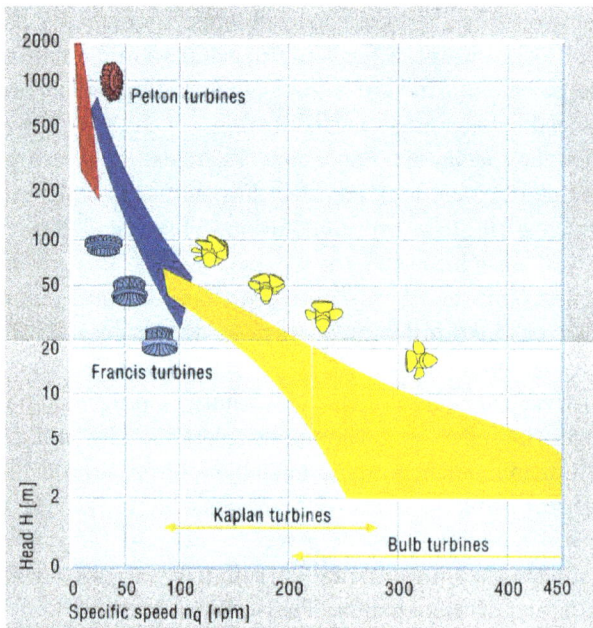

Fig. 37.   Operational ranges for different turbines ($n_q = N_s/51, 6$).

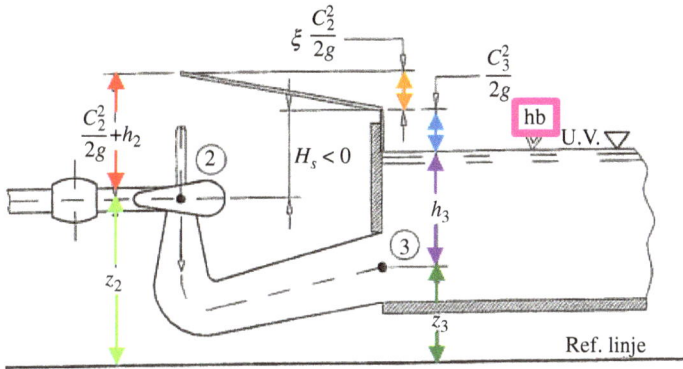

Fig. 38. Schematics used to define properties for cavitation investigations.

becomes so low that the water changes into the gas phase; it is literally boiling, not due to increased temperature, but due to low pressure. These steam bubbles appear as cavities inside the flow, and hence the name Cavitation is appropriate. It is however not this phenomenon that is the destructive one, but the reversed process when the steam bubbles go back to liquid phase. This happens very fast, and a large pressure pulse in induced when the bubbles implode. If this implosion appears close to a solid surface, the materials can be destroyed, each implosion eroding away small pieces of materials. This makes it easy to understand why the cavitation phenomenon must be avoided, so that there are no steam bubbles that can implode. In Fig. 38, a sketch of a hydropower unit is shown, along with the EGL for the flow from the outlet of the turbine to the lower reservoir and energies represented in unit meters. The motivation is to obtain a vertical position for the center of the turbine relative to the water level of the lower reservoir such that the pressure head h_2 is higher than the absolute vapor pressure head h_{va} of the water. By doing this, cavitation will not occur if the detailed design of the runner is not too steeply curved somewhere.

The EGL level at the outlet of the turbine (combined length of red and green arrows up to the upper red arrowhead) can be described by both properties at the turbine (left-hand side of Eq. 17), as well as properties at the outlet of the draft tube and losses between the two positions (right-hand side of Eq. 17)

$$h_2 + \frac{C_2^2}{2g} + z_2 = \xi \frac{C_2^2}{2g} + \frac{C_3^2}{2g} + h_3 + z_3 + h_b \qquad (17)$$

The height H_s in the figure is the one important quantity to find and which is used when constructing the power plant. This can be described by other terms, and introducing the vapor pressure and the requirement that the turbine outlet pressure should be higher than this, we get an inequality describing H_s:

$$H_s < h_b - h_{va} - \left(\frac{C_2^2}{2g} - J \right)$$

where J are the losses from turbine outlet to draft tube outlet. The bracket term is also known as the Required Net Positive Suction Head (NPSH$_R$). This can be difficult to determine, but often the Thoma cavitation coefficient, σ, is used:

$$H_s < h_b - h_{va} - \sigma H_e$$

where values for the cavitation coefficient can be found using empirical relations and are presented in Fig. 39.

One thing to remember is that the highest local velocities and thus lowest local pressures are at the maximum power production. Therefore, this will be the operating conditions for the determination of the submergence to avoid cavitation over the entire range of operation.

If design and calculations are performed in a correct manner, cavitation should not be a problem. But if for some reason cavitation is occurring,

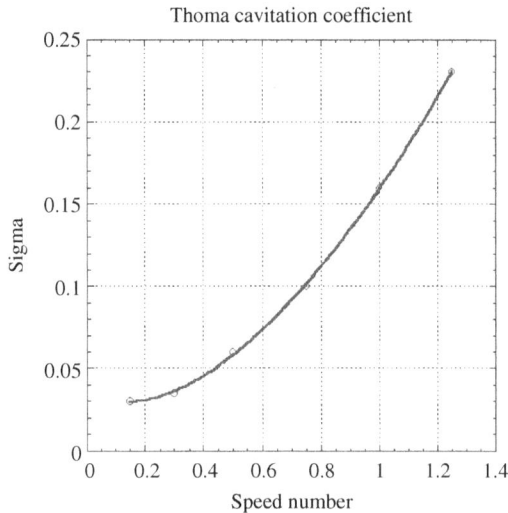

Fig. 39.   Thoma cavitation coefficient (sigma) as function of speed number (= Ns/89).

one thing that can be done to help the situation is to reduce the cross-section area of the outlet to the lower reservoir. This will increase the velocity of the flow at this location and increase the losses in the system. Pressure at the turbine outlet is also increased and this might be enough for avoiding cavitation.

5 Hydropower Services to the Electrical Grid

An electrical grid is the transportation highway for electrons that carry energy. The rate of energy transport is called power, and for an electrical alternating current (AC) grid to operate at stable conditions, there must be a balance between the power consumption and power production. What characterizes a stable AC grid is a constant frequency; in some parts of the world 50 Hz, in other parts 60 Hz. Whenever there is an imbalance, the frequency changes, and the rate of change is dependent on the so-called rotating inertia in the system. The units producing power to the grid have large rotating components that represent stored energy that slows down the rate of change of the frequency so that other measures can be taken to obtain a new balance between consumption and production. It is however the units, operating using *synchronous generators*, that have this stabilizing capability, such as thermal and hydropower units. Solar PV and Wind turbines do not, and a system consisting mostly of such energy producers will be difficult to keep stable. The requirements on frequency are accompanied by requirements on other properties in the grid to obtain a full set of requirements that define a stable high-quality grid operation. These properties are typically the voltage level and the amount of reactive power in the grid.

To be able to maintain a stable situation at all times, there must be power available for all consumption scenarios, which will differ between seasons, months, weeks and hours. The availability of this power is dependent on having enough energy to generate this power, and these longer time-scale balancing objectives are typically referred to as "energy storage". A typical differentiation between these objectives can be seen in Fig. 19.

Depending on the type (run-of-river or storage power plant), hydropower can provide services on a second to seasonal, even annual, timescales. There are however several additional needs that exist in an electrical grid, as can be seen in Table 3. A detailed description of hydropower's capabilities regarding all these services are beyond the scope of this text, but the cells that are shaded represent needs where hydropower plants can provide services.

Table 3. Overview of Energy Storage Applications in the Electricity Sector.[3]

Generation/ Bulk Services	Ancillary Services	Transmission Infrastruc- ture Services	Distribution Infrastruc- ture Services	Customer Energy Management Services
Arbitrage	Primary frequency control	Transmission investment deferral	Capacity support	End-user peak shaving
Electric supply capacity	Secondary frequency control	Angular stability	Contingency grid support	Time-of-use energy cost management
Support to conven- tional generation	Tertiary frequency control	Transmission support	Distribution investment deferral	Particular requirements in power quality
Ancillary services RES support	Frequency stability of the system	—	Distribution power quality	Maximizing self-production & self-consumption of electricity
Capacity firming	Black start	—	Dynamic, local voltage control	Demand charge management
Curtailment minimiza- tion	Voltage support	—	Intentional islanding	Continuity of energy supply
Limitation of Distur- bances	New ancillary services	—	Limitation of disturbances	Limitation of upstream Disturbances
—	—	—	Reactive power compensation	Reactive power compensation
—	—	—	—	EV integration

6 Environmental Impacts and Measures to Avoid Them

Despite hydropower being a renewable source of energy, it is not without environmental impacts. Discussing these impacts, it is convenient to separate local/regional impacts and global impacts. The global impacts are most important when talking about climate change. The renewable energy from hydropower is a sustainable alternative to fossil fuels, and given the high Energy Pay-back Ratio of hydropower (170–280)[b] it will

---

[b] "Energy payback is the ratio of total energy produced during a system's normal lifespan, divided by the energy required to build, maintain and fuel it"; http://tc4.iec.ch/FactSheetPayback.pdf.

give a significant net contribution to a sustainable electrical energy system. For developing countries that are in a situation where they might leap-frog the fossil-driven growth of their economy, hydropower should represent a technology enabling this where available. Some challenges still exist before all hydropower production can be classified as emission free. At some locations, the geographical and thermal conditions will give rise to GHG emissions from decomposing biological matter in the reservoirs, resulting in CO_2 and methane gas release. However, the net effect should not be large, as phytoplankton in the reservoir has a high CO_2 uptake, and for an investigated reservoir in Laos it turned out that this latter effect was greater than the former, and the reservoir had a negative GHG emission,[9] acting as a carbon sink.

Looking at the local/regional impacts, it is important to remember that at many locations the primary objective for constructing a dam has not been to produce power. The primary objective has been to provide water management fulfilling some societal need. The needs have typically been provision of flood and drought control, drinking water, means of transportation, recreation, even to provide cooling water for thermal power plants, or a combination of these. And since the dam has been constructed and some water must flow past the dam, one might as well construct a power plant and get some energy out of this flow. Independent of the main objectives of the dam construction, the consequences on the local environment is typically the same, with or without a power plant installed. The major impacts on biodiversity are habitat loss on land and obstruction of animal migration routes due to the reservoirs, obstruction of fish migratory paths and subsequent fragmentation of populations. There are also effects on ecology due to alterations in flow and temperature regimes of the river, and effects due to alterations in sediment and organic matter transport.[10]

6.1 *Mitigation measures*

Mitigation measures of hydropower projects can be classified into 10 groups of biophysical and socio-economic classes:

> *Reservoir impoundment (Fisheries and Terrestrial Habitats); Loss of biological diversity; Reservoir sedimentation; Modifications to water quality; Modifications to hydrological regimes; Barriers for fish migration and river navigation; Involuntary displacement; Public health risks; Impacts on vulnerable minority groups; Sharing of development benefits.*

The mitigation measures for each of these can be divided into four categories, depending on what phase of the execution they are implemented[11]:

- Impact avoidance measures: Usually implemented in the design- and planning stage.
- Mitigation measures: Used to eliminate or reduce the intensity of a source of impact.
- Compensation measures: Seeking to compensate impacts or residuals of such after mitigation measures are implemented.
- Enhancement measures: Improving existing environmental or social conditions which are not directly affected by the project.

As an example, the *Barrier for fish migration and river navigation* lists different techniques most effective for upstream and downstream movement:

Upstream:

- Locks, lifts and elevators for watercraft.
- Fishways, bypass channels, fish elevators, with attraction flow or leaders to guide fish to fishways.
- Capture and transportation of fish upstream.

Downstream:

- Improvement in turbine, spillway or overflow design.
- Management of flow regime or spillway during downstream movement of migratory fish.
- Installation of avoidance systems upstream of the power plant.
- Capture and transportation of fish downstream.

An extensive description of all techniques for mitigation of impact corresponding to the remaining nine classes is beyond the scope of this text, but readers interested in this part of hydropower projects are encouraged to read the referenced paper.[11] As any hydropower project is unique, all techniques are on the table initially in the project. A detailed analysis of a project must be performed in order to find the techniques that are likely to be effective in that project. As a concluding remark, the requirements for environmental friendliness in hydropower projects are becoming more and more important, which is as it should be.

References

1. IHA (2019). 2019 Hydropower status report.
2. G. Ventrone (2002). Macchine per allievi ingegneri. Padova, Italy: Cortina.
3. European Association for Storage of Energy (2014). EASE Activity report. http://ease-storage.eu/wp-content/uploads/2015/10/EASE-Activity-Report-2014_LR.pdf.
4. N. S. Pearre and L. G. Swan (2015). Technoeconomic feasibility of grid storage: Mapping electrical services and energy storage technologies. *Applied Energy*, **137**, 501–510.
5. K. Bradbury, L. Pratson, and D. Patiño-Echeverri (2014). Economic viability of energy storage systems based on price arbitrage potential in real-time U.S. electricity markets. *Applied Energy*, **114**, 512–519.
6. EASE-EERA (2017). European energy storage technology development roadmap — 2017 Update.
7. J. I. Pérez-Díaz, G. Cavazzini, F. Blázquez, C. Platero, J. Fraile-Ardanuy, J.A. Sánchez, and M. Chazarra (2014). Technological developments for pumped-hydro energy storage. Technical report, mechanical storage subprogramme, joint programme on energy storage, *EERA*.
8. Y. Cencel and J. M. Cimbala (2013). Fluid mechanics fundamentals and applications. New York, USA: McGraw-Hill Higher Education.
9. V. Chanudet, S. Descloux, A. Harby, H. Sundt, B. H. Hansen, and O. G. Brakstad (2011). Gross CO_2 and CH_4 emissions from the Nam Ngum and Nam Leuk sub-tropical reservoirs in Lao PDR. *Science of The Total Environment*, **409**(24), 5382–5391.
10. E. O. Gracey and F. Verones (2016). Impacts from hydropower production on biodiversity in an LCA framework — Review and recommendations. *The International Journal of Life Cycle Assessment*, **21**, 412–428.
11. S. Trussart, Messier, V. Roquet, and S. Aki (2002). Hydropower projects: A review of most efficient mitigation measures. *Energy Policy*, **30**(14), 1251–1259.

Chapter 6

Ocean Thermal Energy Conversion (OTEC)

Gérard C. Nihous

Department of Ocean and Resources Engineering,
University of Hawaii Honolulu, Hawaii, USA
nihous@hawaii.edu

Abstract
Stable temperature differences of the order of 20°C between the surface
of most tropical oceans and water depths of about 1 km can be used to
drive a heat engine. Formulated in the late 19th Century, the concept
is known as Ocean Thermal Energy Conversion (OTEC). While deep
water marine technologies are challenging *per se*, the small temperature
difference available to OTEC systems results in the need for large components and high capital costs. OTEC was first tested at sea by Georges
Claude in the 1920s. Further research and development took place in the
late 20th Century, mostly in the US, but this work fell short of establishing OTEC on a firm commercial footing. While renewed efforts are
under way to deploy and operate floating OTEC pilots of 5–10 MW, this
vast renewable resource with exceptional base-load capabilities remains
untapped.

1 Basic Concept of Ocean Thermal Energy Conversion

Wherever seawater temperature varies sharply with depth and the stratification of the water column is stable, it is theoretically possible to extract
mechanical power in a heat engine. This basic concept of Ocean Thermal
Energy Conversion (OTEC) was first articulated in 1881 by d'Arsonval,[1]
although several decades would pass before significant field tests of OTEC
would take place.

Ocean temperature differences of the order of 20°C can be found within
the upper kilometer of most tropical oceans, between latitudes 30°N and

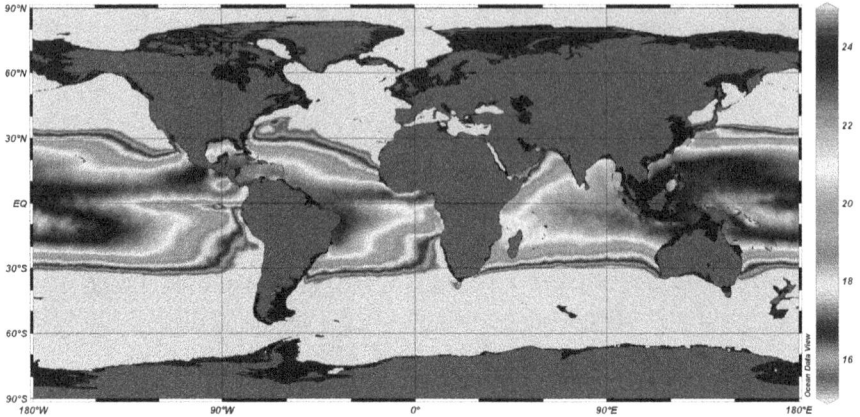

Fig. 1. Yearly average seawater temperature difference between 20 m and 1000 m water depths (°C — grayscale palette between 15 and 25).[2]

30°S. This is illustrated in Fig. 1. Strong cold currents and upwellings along the western coasts of Africa and America locally reduce the zonal extent of the OTEC thermal resource. Other specific regional factors like the Red Sea outflow also affect the availability of vertical temperature differences exceeding 20°C. Yet, the area of interest for OTEC extends over 100 to 120 million km.[2]

The existence of steep stable thermoclines in tropical oceans may be taken for granted but is far from obvious. That intense solar radiation would warm up the surface layer is clear. In the absence of another fundamental physical process, however, the downward diffusion of heat would tend to homogenize the water column, as happens in the great lakes of Africa. Instead, a vast network of planetary currents is responsible for supplying deep cold seawater of polar origin throughout the world's oceans. Technically known as the thermohaline circulation, this Great Ocean Conveyor Belt is sketched in Fig. 2.

The heat engine envisioned by d'Arsonval is similar to any machine that cyclically receives heat from a hot reservoir at temperature T_1 and rejects heat to a cold reservoir at temperature T_2. In other words, the fundamental thermodynamic description of an OTEC plant is similar to that of other thermal power plants. In the simplest version of OTEC, an auxiliary working fluid like ammonia follows a Rankine cycle in a closed loop schematically depicted in Fig. 3. Between Points 1 and 2, the working fluid warms up and boils in an evaporator fed by warm surface seawater; it

Thermohaline Circulation

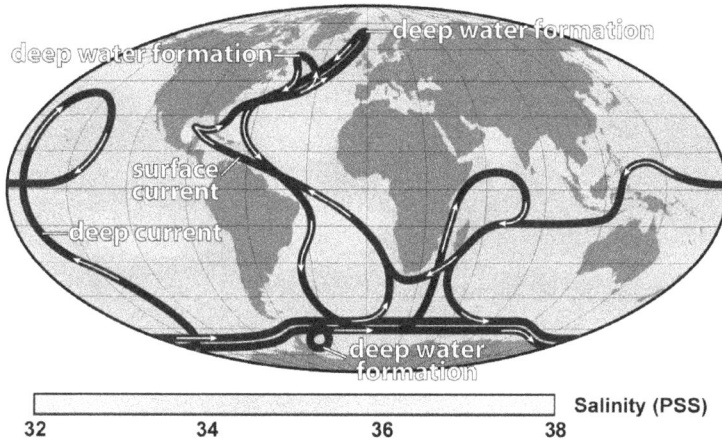

Fig. 2.   Sketch of the thermohaline circulation.[3]

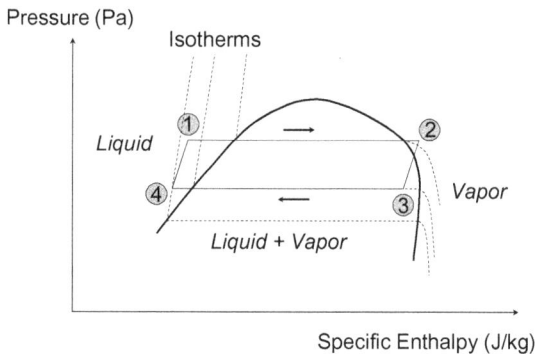

Fig. 3.   Schematic diagram of a closed-cycle Rankine process.

then expands through a turbine between Points 2 and 3, where mechanical work is produced; the vapor leaving the turbine enters a condenser fed by deep cold seawater where it condenses to a liquid and cools to the state represented by Point 4. The cycle is completed by pumping the liquid from Point 4 back to Point 1.

A simplified heat-and-mass balance for a typical OTEC plant that would produce about 10 MW of net electrical power is shown in Fig. 4. The working-fluid closed loop is drawn in gray. It is worth noting that the mechanical power consumed by the working fluid pump is negligible

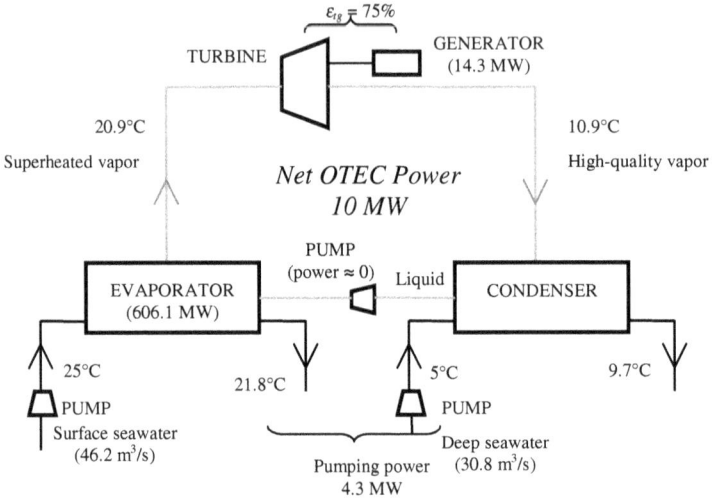

Fig. 4.   Simplified heat-and-mass balance of a typical OTEC plant.

when compared to the mechanical power produced by the turbine. Using the Clausius–Clapeyron equation, it can be shown that the ratio of these terms is approximately equal to the working fluid vapor density divided by the working fluid liquid density.

Although OTEC systems are not conceptually different from other thermal power plants, in-plant power consumption is much larger, with values of the order of 30% of the turbo-generator output. These parasitic pumping power losses P_P occur *outside* of the working fluid loop, as a result of the significant seawater flow rates needed in the heat exchangers. Because surface seawater is more accessible than cold seawater drawn from depths of about 1000 m, plant optimization generally leads to a higher surface seawater flow rate.

If one considers the seawater temperatures available to an OTEC plant, they are the same as those involved in room-temperature refrigeration. In this case, however, mechanical power is consumed in a compressor (driven by an electric motor) to artificially maintain a temperature difference that would not exist otherwise. Hence, the working fluid loop in Fig. 4 essentially describes a refrigerator that would be run in reverse. Not surprisingly, the same substances that are used in moderate refrigeration are adequate for OTEC. From an engineering point of view, these fluids have high enthalpies of condensation per unit volume of vapor in the temperature range of interest, which is equivalent to having steep saturation curves. With excellent heat transfer characteristics to boot, ammonia is an excellent choice.

Heat engines are often evaluated by their thermodynamic efficiency η, i.e. the ratio of power produced by the working fluid through the cycle divided by the heat flow rate from the hot reservoir. By considering an engine operating reversibly between two reservoirs at absolute temperatures T_1 (higher) and T_2 (lower), Carnot established the upper limit $\eta_C = 1 - T_2/T_1$. Such an ideal machine, however, would not produce any power because reversible heat exchange would (have to) be exceedingly slow. An analysis of the maximum output of semi-ideal (or endoreversible) engines, which exchange heat irreversibly at constant temperature but operate reversibly otherwise, yields the surprisingly elegant result[4] $\eta_E = 1 - \sqrt{T_2}/\sqrt{T_1}$. This is applicable to a Rankine cycle because heat exchange at constant pressure *also* takes place, for the most part, at constant temperature as the working fluid changes phase (e.g. between Points 1 and 2, or Points 3 and 4 in Figs. 3 and 4). In the particular case of OTEC, $T_1 \approx T_2$ (in Kelvin) so that we have $\eta_E \approx 1 - T_2/(2T_1)$. In other words, only half of the available seawater temperature difference $(T_1 - T_2)$ should be used across the turbine, while the rest would allow the optimal transfer of heat. This is graphically illustrated by the temperature ladder shown in Fig. 5. The analysis of semi-ideal machines also allows an estimation of the terms x and y, as well as of the cycle (endoreversible) power P_E.

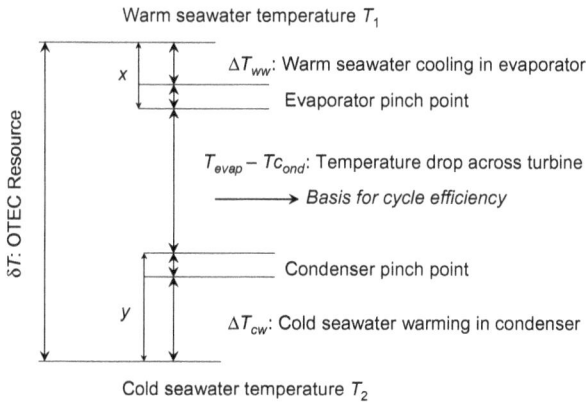

Fig. 5.   OTEC temperature ladder.

The latter is proportional to $(\sqrt{T_1} - \sqrt{T_2})^2$, which for OTEC conditions is about $(T_1 - T_2)^2/(4T_1)$.

To accurately describe OTEC systems, additional irreversibilities have to be taken into account. The turbine and working fluid pump are not ideal isentropic machines, and the conversion of mechanical power to electrical power is imperfect; this can be quantified by specific efficiencies. Moreover, working fluid pressure drops in the heat exchangers should be minimized since they will diminish the actual pressure difference across the turbine.

With seawater temperature differences $(T_1 - T_2)$ of the order of 20–24°C, values of η_E (3–4%) are about ten times smaller than the thermodynamic efficiencies of thermal power plants that produce electricity from fossil or nuclear fuels. While much has been made of this fact *per se*, its meaning should be interpreted cautiously because OTEC resources are abundant and renewable. The limited thermal potential difference available to OTEC power plants does affect their design in many ways, however. This will be discussed in Section 3.

An immediate consequence of the dependence of P_E on the square of $(T_1 - T_2)$ is the sensitivity of power output on available seawater temperatures. Namely, a change of 1°C in $(T_1 - T_2)$ will correspond to a variation of about 10% for P_E or P_G, the gross power produced by the turbogenerator (which is essentially P_E multiplied by turbine and generator efficiencies). This 10% rule of thumb is illustrated in Fig. 6 where time histories of P_G and $(T_1 - T_2)$ are displayed for an experimental land-based OTEC plant operated through the early 1990s in Hawaii; in this particular instance, a large warm eddy moving along the coast rapidly affected T_1 and, as a result, P_G.

Matters are actually more acute for the *net* power $P_{net} = P_G - P_P$ produced by OTEC plants, because the parasitic seawater pumping power P_P remains essentially the same when $(T_1 - T_2)$ varies (there is only a small margin for operational optimization). With P_P about 30% of P_G, a change of 1°C in $(T_1 - T_2)$ will correspond to a variation of about 15% for P_{net}.

Finally, a unique OTEC cycle which does not resemble a refrigeration system in reverse is briefly discussed here. The Rankine cycle is closed, with the working fluid moving in a sealed loop. In the 1920s, the brilliant engineer and successful entrepreneur Georges Claude conceived instead a process known as Open-Cycle OTEC, where the working fluid consists of water vapor continuously produced in the evaporator and consumed in the

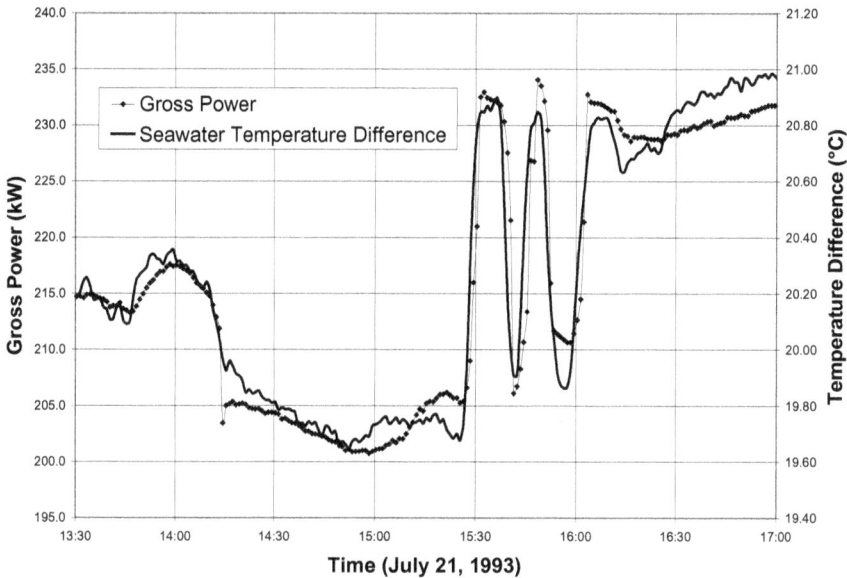

Fig. 6. Sample time histories of gross power and available seawater temperature difference from an experimental OTEC plant.[5]

condenser after its passage through a turbine.[6] In Figs. 3 and 4, this would correspond to a removal of the step between Points 4 and 1; otherwise, the thermodynamic analysis of OC-OTEC is not fundamentally different. The boiling of about 0.5% of the surface seawater in a flash evaporator relies on very low pressures of less than 3 kPa throughout the system. With an Open Cycle (OC) plant, large and costly metal heat exchangers are replaced by simple vacuum chambers, unless fresh water production in a standard condenser is desirable. This also reduces or eliminates the need for seawater treatment to prevent biofouling, the buildup of bacterial mats on heat conducting metal surfaces.[7] Such benefits are offset by the need for large low-pressure turbines as well as for vacuum compressors. The latter must continuously remove the non-condensable gases released from seawater exposed to very low pressures. Owing to the perseverance and wealth of Claude, OC-OTEC was first tested in the field in the late 1920s. It is also the OC-OTEC system shown in Fig. 7 that was last operated in the 1990s after setting a number of world records[8] (power production, length of operation, grid connection and desalination potential); the data in Fig. 6 were generated by this experimental plant.

Fig. 7. Aerial view of an experimental OC-OTEC plant in Kailua-Kona, Hawaii; the heat exchangers and turbine were enclosed inside the concrete cylindrical vacuum structure in the lower left corner.

2   Available OTEC Resources

In the absence of widespread OTEC development, the question of an accurate estimation of worldwide OTEC resources may seem rather academic, although there is also a regional aspect to it (local maximum OTEC power production density). In the past, proponents of OTEC technologies have routinely used the enormous amount of solar power absorbed by the upper layer of tropical oceans as a starting point to evaluate OTEC resources. Not surprisingly, this leads to the conclusion that when measured against mankind's current total energy use of 1.4×10^{14} kWh per year (16 TW), OTEC resources are virtually unlimited in spite of low thermodynamic efficiencies.

A more sober approach consists in viewing OTEC as a true energy conversion technology. The thermal potentials provided by the temperature stratification of the water column in tropical regions are very much like elevation differences in a hydropower plant. What is not obvious for OTEC is an analog for the river flow rate that also is a fundamental factor in determining hydropower resources. The driving mechanism for the existence of a stable pool of deep cold seawater everywhere is the planetary thermohaline circulation depicted in Fig. 2. It has been suggested that the rate of deep water formation in polar areas is a natural flow rate scale for sustainable OTEC power extraction on a massive scale. Based on a cold seawater flow rate intensity of $3\,m^3/s$ per net OTEC megawatt and a thermohaline circulation of the order of 30 Sv (1 Sv = 1 million m^3/s), Cousteau and Jacquier[9] argued that OTEC resources were as large as 10 TW.

Because the seawater needs of OTEC systems are great, however, simple one-dimensional analyses of the water column with OTEC showed that it was theoretically possible to reach a point where the available vertical temperature difference itself could be eroded. In such scenarios, overall OTEC power actually reaches a maximum of about 3 to 5 TW when the combined OTEC cold seawater demand is roughly equal to the rate of deep water formation.[10] Such one-dimensional models were known to be too conservative because the great geographic extent of the area favorable for OTEC only represents about a third of the overall oceanic surface. Therefore, horizontal diffusive and advective transport mechanisms that are inherently much stronger in the ocean could not be properly accounted for.

The use of better models that include three-dimensional ocean dynamics and a detailed heat budget at the ocean–atmosphere interface was initiated only recently. This numerically intense effort to better assess global OTEC resources has been described in a series of articles, from the original study with a limited grid resolution[11] (4° by 4° horizontally, 15 vertical ocean layers) to the next[12] (1° by 1° horizontally, 23 vertical ocean layers). The issue of broad regional constraints on OTEC, for example, by limiting its development to the EEZ, or within 100 km of coastlines, was then considered.[13] Finally, a low-complexity atmospheric model was coupled with the ocean general circulation model to better represent feedback processes between the ocean and the atmosphere.[14]

Significant and relatively robust results were obtained. On the one hand, the existence of large-scale maxima for OTEC net power production was confirmed, when the seawater flow rates needed to sustain OTEC plants eventually degrade the available thermal resource; as expected, such maxima were predicted to be greater than in simplified one-dimensional analyses. Figure 8 illustrates this trend with output from the most advanced version of the modeling suite.[14] The abscissa refers to net OTEC power that would be produced *in the absence of environmental feedback.*

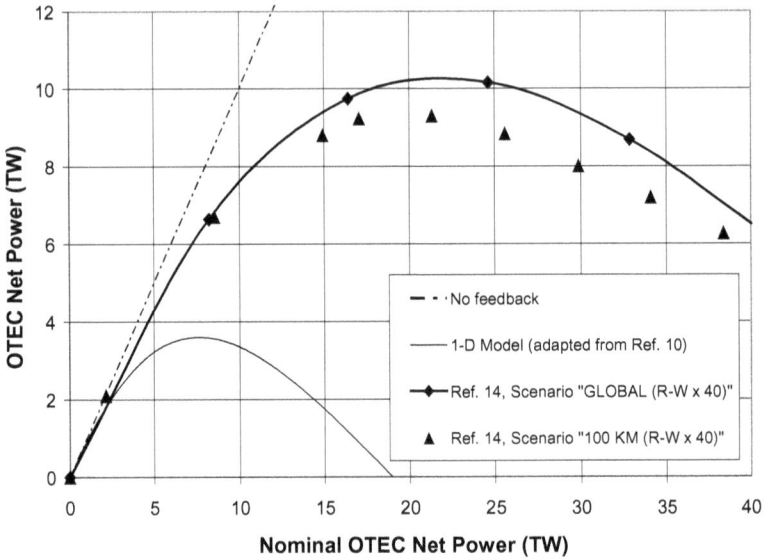

Fig. 8.   Estimates of OTEC net power as a function of nominal (no-feedback) values in 1-D and 3-D models.[10, 14]

Figure 8 also shows that restricting the development of OTEC to the vicinity of coastlines, nominally within practical reach of submarine power cables (∼100 km) hardly affects the outcome; such a counter-intuitive result means that a greater density of OTEC plants can be envisioned in a geographically restricted region; in this scenario, however, the local thermal resource essentially undergoes the same degradation, which highlights the dominance of the horizontal heat transport phenomena.

All model results share the fact that actual OTEC net power maxima only represents 40–50% of the corresponding nominal values (same flow rates with no interactions). Maxima roughly correspond to 0.2 TW per Sverdrup of OTEC cold seawater, or equivalently, it would take as much as $5 \, \mathrm{m^3/s}$ of deep cold seawater to produce, on average, 1 MW.

More importantly, output based on an ocean general circulation model reveals that profound environmental changes would occur if OTEC were implemented at maximum net power capacity (∼10 TW). On the one hand, a very large amount of heat would initially be transferred into the ocean interior, until heat exchange at the ocean–atmosphere boundary reaches a new equilibrium. Such heat would also be redistributed well outside the OTEC area, with a surface cooling trend in the tropics balanced by surface warming in some high-latitude regions. This teleconnection would result from the propagation of long waves in the ocean interior, as well as from significant modifications of large-scale ocean currents. Hence, it would be prudent to recognize that any OTEC development toward a stable power production limit for this technology might actually correspond to excessive or undesirable environmental effects.

Background uncertainties related to global warming would add significant complexity to the problem, even though the time scale of any widespread implementation of OTEC would, in all likelihood, exceed that of current climate change trends. Until better modeling approaches are available, it is safe to say that OTEC is a terawatt-size global resource.

3 Advantages and Disadvantages of OTEC

As suggested by Fig. 1 and following the discussion in Section 2, OTEC represents a vast potential contributor of renewable power. However, the fact that OTEC development has not proceeded beyond the pre-commercial stage warrants some examination.

Offhand, it is clear that marine technologies are more difficult to implement than their land-based counterparts or competitors. This was clearly demonstrated by the history of offshore oil and gas exploration and production, which only took off in the 1970s. Such a development was spurred by OPEC's decision to sharply increase the price of crude oil. More recently, the commercial success enjoyed by the wind power industry also shows that a move from land-based systems to shallow offshore farms has been both difficult and costly. In fact, OTEC not only is a marine renewable technology, but it requires deep waters where even wind farms have yet to be deployed. The remoteness of offshore OTEC plants offers, of course, a few conceptual benefits. There would be no interference with land and coastal communities and ecosystems.

Another unique attribute of OTEC is its geopolitical distribution in tropical regions. Although this point is seldom mentioned, it is believed to represent a strong disadvantage. Essentially located across higher latitudes, the wealthiest and most technologically advanced nations may be reluctant to commit their taxpayers' contributions toward developing scarce domestic OTEC resources.

From an engineering point of view, the OTEC zone includes the area where powerful tropical cyclones occur since OTEC and these storms share a need for warm surface seawater. Hence, the design of offshore OTEC systems would have to follow very strict standards. OTEC resources that are safe from hurricanes and typhoons lie along the Equator where the Coriolis force becomes negligible, and in the tropical South Atlantic where high-altitude wind shear prevents the formation of strong cyclones.

A remarkable quality of OTEC among renewable energy technologies is a potential for very high capacity factors. Within the reach of submarine power cables, OTEC would be capable of supplying base-load power. This would be an exceptional advantage for isolated electrical grids. It was mentioned in Section 1 that OTEC power output is very sensitive to available seawater temperatures. From the point of view of an electrical utility, the definition of base-load power might therefore hinge on some minimum expected OTEC power output. In this respect, a map of yearly average seawater temperature differences like Fig. 1 is not sufficient because it does not show seasonal changes of the thermal resource. Fortunately, a very large fraction of the region deemed favorable for OTEC in an average sense also exhibits low seasonal variability (e.g. within 3 to 4°C). In Fig. 9, historical monthly seawater temperature differences at either an excellent site, off of Fortaleza (Brazil), or at an unfavorable site, off of

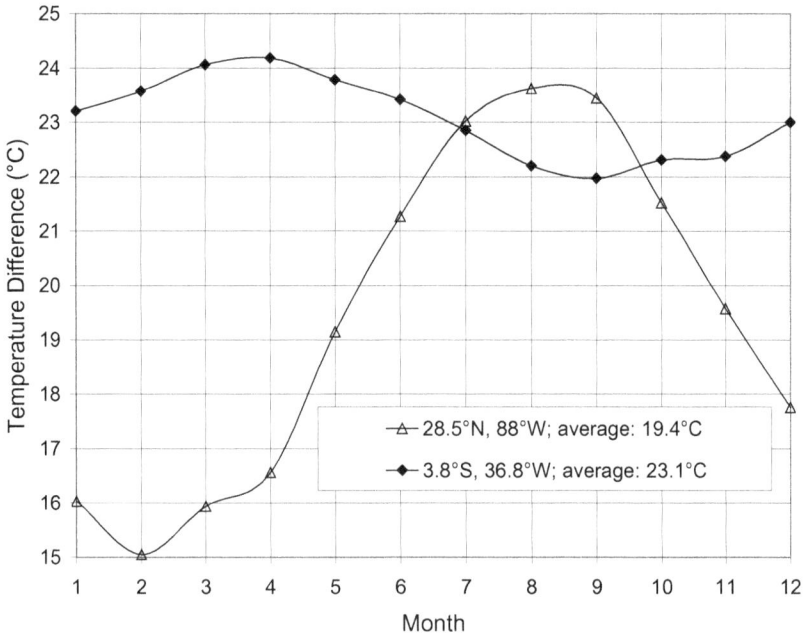

Fig. 9.   Examples of monthly seawater temperature differences between 20 m and 1000 m depths; data derived from the 2005 World Ocean Atlas.[15]

Mobile (US Gulf Coast), show the need to assess variability; designing practical or cost-effective OTEC systems in the northern Gulf of Mexico would be very difficult.

Beyond the practical reach of large submarine power cables, the development of OTEC resources would have to rely on the *in situ* manufacture of liquid fuels. Typically, hydrogen could be produced from electrolysis, and stored and transported as such, or used to manufacture ammonia. Taking into account power losses and process efficiencies, a commercial OTEC plant ship producing 100 MW of electrical power could manufacture 30 metric tons of hydrogen per day (delivered to shore). Given the energy density of hydrogen, this would be equivalent to about 750 barrels of oil per day. Though substantial, this figure may be put in perspective by comparison with the 1999 peak North Sea oil production of 6 million barrels per day.[16]

The most definite consequence of the small temperature differences that characterize OTEC processes is the need for large systems. Heat exchanger surfaces of the order $10000 \, \text{m}^2$ per net MW typically are required either for the evaporator or for the condenser. High seawater flow rate intensities are necessary as well, with cold seawater needs of 2.5 to $3 \, \text{m}^3/\text{s}$ per MW and somewhat larger warm seawater flow rates (as a result of better accessibility). Since the cooling fluid is drawn from depths of about 1000 m, the size of the cold water pipe (CWP) presents a substantial technological challenge. Resorting to high pumping velocities is not a viable option because parasitic losses would rapidly reach unacceptable levels; for a given flow rate, friction losses, for example, depend inversely on the fifth power of the pipe diameter. Therefore, one has to consider CWP diameters as large as 12 m for a 100 MW OTEC plant; in truth, this is what limits the probable size of a single OTEC system.

The size of OTEC hardware and its offshore location result in high capital costs. From available design studies and best estimates, a steep economy of scale is also expected since several critical components are not modular and there are some substantial fixed costs involved as well. This point is clearly illustrated in Fig. 10 where capital costs per unit kW drop sharply; here, OTEC power production is assumed to take place under some standard conditions (e.g. temperature difference of 22°C). Figure 11 shows the cost breakdown for a typical commercial 100 MW unit; the overall investment burden would be $750 million. The most expensive items consist of the heat exchangers and the floating vessel, although the CWP and its installation present the highest risk and would define a technological

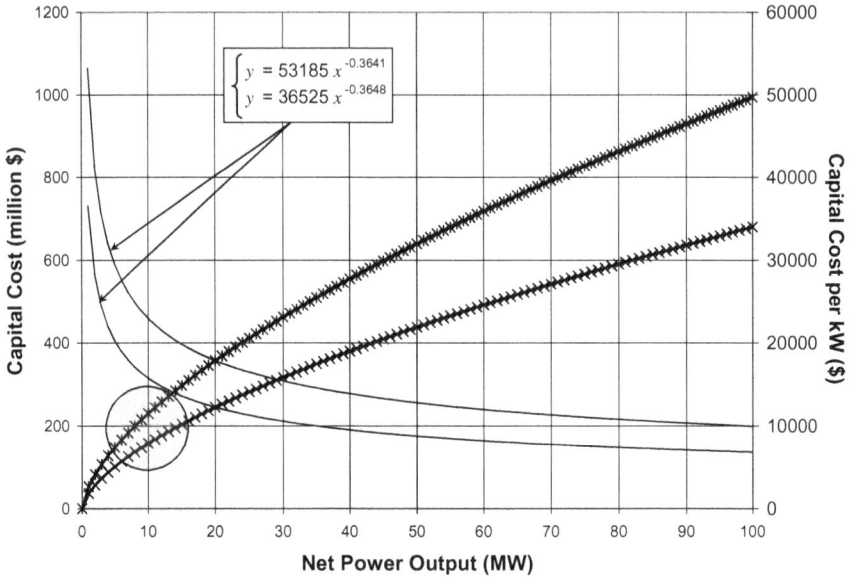

Fig. 10. Capital cost estimates for OTEC plants;[17] the smooth lines define a range for cost per kW; the circle suggests desirable pre-commercial pilot projects.

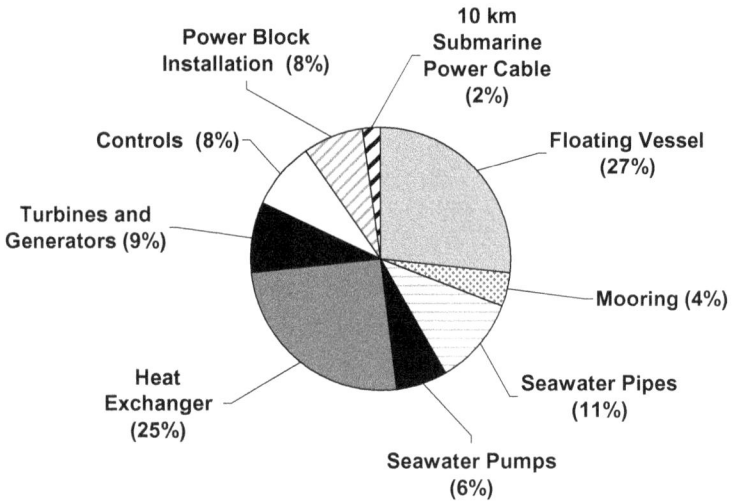

Fig. 11. Estimated capital cost breakdown for a 100 MW commercial OTEC system (overall cost of $750 million).[17]

frontier even today. Submarine power cable costs depend almost linearly on distance, while the influence of power carrying capacity is weaker. For a commercial OTEC plant or a pilot, installed cable costs of $1700/m and $1300/m, respectively, are typical. Hence, the relative burden imposed by power transmission to shore for a plant located 100 km at sea would rise to nearly 20% of a significantly higher investment.

OTEC by-products have received much attention as well, in attempts to boost projected revenues when estimating cost effectiveness. The best known is desalinated water for OC-OTEC. Roughly speaking, a fresh water output of 23 kg/s per net MW could be generated in a typical OC-OTEC system; using the residual temperature difference of the plant's seawater effluents in an additional (second-stage) desalination module could double that figure (note that the residual thermal resource would be too small for additional electricity generation). Therefore, a single-stage 100 MW commercial OC-OTEC plant could produce about 200000 m^3/day. This is a significant amount, but it must ultimately be evaluated by taking into account the market value of fresh water and the need for storage and transport from offshore OTEC plants.

Other touted by-products hinge on the specific properties of deep seawater alone. Cold, clean and nutrient-rich deep seawater can be used in air-conditioning applications, aquaculture operations and the production of high-value items like nutraceuticals. There have been a few successful implementations of coastal seawater air-conditioning (SWAC) systems, and many technology incubators patterned after the *Natural Energy Laboratory of Hawaii Authority* (NELHA) at Keahole Point, Hawaii, have been built around the world, notably in Japan. Generally speaking, however, there is a disconnect between the coastal nature of most by-product technologies and the offshore character of commercial-scale OTEC power production, as well as a mismatch between the great seawater flow rate intensity of OTEC systems and the seawater needs of by-product technologies.

The only idea that fully relies on massive offshore seawater flow rates is artificial upwelling (AU): deep-seawater nutrients (nitrates, phosphates, etc.) brought to the surface for OTEC and released in the plant's effluents could spur phytoplankton growth and boost the marine food web if the discharge plume settled within the photic zone, where sufficient solar radiation is available. AU is conceptually possible but practically untested. For lack of better knowledge, existing permitting frameworks for OTEC projects are likely to adopt a strict cautionary principle that would aim to prevent potential environmental effects from OTEC effluents on the marine environment. From a wider perspective, the overall legal context that would apply to the large-scale development of offshore OTEC resources remains uncertain.

4   History and Status of OTEC Development

The history of OTEC development essentially reflects the technological and financial challenges at the heart of this technology. It globally proceeded through two distinct periods where a few ambitious field projects were attempted or completed. A summary description is provided in what follows, while more details can be found elsewhere.[18]

A pioneering phase from 1926 to 1935 corresponds to the single-handed efforts of Georges Claude. This exceptionally creative chemical engineer developed and patented a number of breakthrough industrial processes (acetylene storage, air liquefaction and neon lighting), but also proved to be an accomplished business leader when he founded Air Liquide with Paul Delorme in 1902. Such undertakings secured him great wealth and established him as a charismatic figure.[19, 20] Convinced of the enormous potential of OTEC and determined to spearhead its development, he was also keenly aware of the practical difficulties involved, and of the need to convince potential investors. His Open Cycle concept was first tested in 1928 with a small plant built along the Meuse River at Ougrée (Belgium), where warm effluents from a steel mill and cold river water provided a 20°C temperature difference similar to ocean OTEC conditions. This pilot system produced a gross power of 60 kW as expected, and Claude transferred it to the shores of Matanzas Bay, in Cuba. There, for nearly two years, Claude faced the daunting technical problem of deploying a CWP 1.6 m in diameter and 2 km long over the seafloor to a water depth of 700 m. The conduit consisted of corrugated steel sections connected by rubber joints, and it was deliberately oversized for fear of excessive warming of the cold seawater (see Fig. 12). After two spectacular failures where the assembly of the CWP from floating sections had been attempted at sea or in a nearby stream, Claude was left alone to finance his OTEC projects. The Cuban engineer Garcia Vazquez then conceived a new CWP deployment method where the conduit would initially be assembled on a small railway perpendicular to the shoreline, pulled out to sea afloat by a tugboat, and finally deployed to the seafloor after controlled immersion. Even though such an elegant solution considerably reduces exposure to the elements, another CWP was lost due to a

Fig. 12.   Section of the 1.6 m diameter CWP being assembled on shore for the OC-OTEC demonstration plant of Matanzas, Cuba in 1930.

field operator miscommunication during the final immersion step. A fourth try eventually met with success, and the Matanzas OTEC plant produced as much as 22 kW in October 1930.

This engineering achievement proved to be costly and bittersweet for Claude,[21] though it certainly proved his tenacity. He faced skepticism from scientific and technical circles, who were not impressed by the performance of the plant in spite of overwhelming but quantifiable obstacles (mostly, an oversized CWP with leaky joints that degraded the available thermal resource). Accordingly, investors and politicians continued to show little enthusiasm for OTEC. Always the innovator, Claude turned his attention to floating OTEC systems. After buying and refurbishing the ship *le Tunisie*, he moved it to waters off Rio de Janeiro where the power produced on board would be used to make ice, a commodity in great demand (Fig. 13). The CWP was designed as a vertical 2.5 m diameter, 650 m long conduit connected to a spherical float, to be assembled on site offshore in sections (Fig. 14). A flotilla of 7 boats with a crew of 80 was involved in deployment operations. This eventually led to costs too high for Claude to bear, in the midst of the Great Depression. The project, marred by numerous delays and unfavorable weather, faced an ultimate crisis in early

Fig. 13. Floating OC-OTEC ice factory *Le Tunisie* off Rio de Janeiro.

1935 when the suspended CWP deadweight anchor broke off during pipe assembly.

Following the so-called oil shocks of the 1970s, and with the benefit of four decades of technological advances in the fields of ocean engineering and material science, OTEC saw a renaissance when governments in the US and Japan launched research and development initiatives that led to the execution of several small but high-profile projects in Hawaii and Nauru. In nearly all cases, a critical element for success was the availability of high density polyethylene (HDPE) pipes of moderate diameters, with low bending stiffness. In 1977, a net power production of 20 kW was achieved by *Mini-OTEC*, a small barge fitted with a closed-cycle system using ammonia, although turbine and pumps were not very efficient. The 0.6 m diameter, 670 m long HDPE CWP was connected to a surface buoy rather than directly to the OTEC platform, following the concept articulated by Claude for his *Tunisie* project more than four decades earlier.

Three years later, in 1980, a much larger floating system known as *OTEC-1* was deployed in waters 40 km off the west coast of the island of Hawaii. For this project under the auspices of the US Department of Energy (DOE), a turbogenerator regrettably was excluded, and onboard electrical needs were met with diesel generators. The ammonia heat exchanger rating

Fig. 14.   Assembly of a section of the 2.5 m diameter vertical CWP for the floating OC-OTEC ice factory *Le Tunisie*, 1935.

of nearly 40 MW suggests that *OTEC-1* could have produced net power of about 1 MW. The CWP consisted of a bundle of three 1.2 m diameter HDPE conduits connected to a floatation collar at one end, and a ballasting weight at the other. It was assembled on shore following the method first proposed by Garcia Vazquez (Fig. 15), pulled afloat into Kawaihae harbor, towed to the *OTEC-1* offshore site, and carefully upended before its final connection to a gimbal joint under the plantship.[22] The floating thermal plant was operated for several months as long as operating funds allowed. In spite of criticisms, the project's scope and successful execution remain unsurpassed to date among experimental offshore OTEC systems.

At nearly the same time, in 1981, Tokyo Electric Company and Toshiba Corporation deployed and tested a small land-based OTEC pilot plant on the isolated island of Nauru, in the Equatorial Pacific. The working fluid R-22 was selected at the time, well before concerns about the

Fig. 15. *OTEC-1* CWP bundle of 3 HDPE conduits being assembled onshore at Kawaihae harbor, island of Hawaii (1980).

ozone-depleting characteristics of such refrigerants emerged. A peak gross power production of 120 kW was achieved, and 31.5 kW could be exported to the local electric grid. Continuous operation (24 hours a day) was also demonstrated over a period of 10 days.[23] The seafloor mounted 0.7 m diameter, 945 m long HDPE CWP was damaged in a storm just before the project reached its planned completion.

While small OTEC systems could rely on available HDPE conduits less than 2 m in diameter, the feasibility of larger, stiffer CWPs had yet to be established. To this end, the National Oceanic and Atmospheric Administration (NOAA) of the US Department of Commerce launched an ambitious research program designed to demonstrate that CWPs of the order of 10 m in diameter were within the state of the art. Following theoretical and experimental (model-basin) investigations, an at-sea test for a 1/3 scale fiberglass-reinforced-plastic (FRP) CWP, 2.4 m in diameter and 400 m long, was planned off Honolulu in 1983. A 12 m section being unloaded at Honolulu harbor can be seen in Fig. 16. It had become clear earlier, however, that budgetary constraints imposed by the new Reagan administration in the early 1980s would only allow a shorter 120 m long CWP to

Fig. 16. A 12 m section of the 120 m, 2.4 m diameter FRP CWP tested at sea off Honolulu in 1983 is unloaded at the harbor.

be tested. Although a reduced length would alter the dynamic response of the suspended pipe, motions and deformations remained predictable with available analytical tools. The project team, therefore, opted to proceed, and the at-sea test of the 120 m long FRP pipe hanging from a small barge was successfully conducted in May 1983 for three weeks. Strain-gauge data measured along the CWP were used to verify and calibrate computer predictions of bending stress.

The last major successful OTEC demonstration project to date, known as the *Net Power Producing Experiment* (NPPE), was undertaken in the mid-1980s. The State of Hawaii still manages the selected site today, at the NELHA near the Kona airport, on the west coast of the island of Hawaii. It took responsibility for the financing and installation of all seawater systems (pipes, pumps and sumps), since this infrastructure was intended to serve the needs of other users (e.g. mariculture incubators) during and well beyond NPPE. The centerpiece was a 1 m diameter, 1916 m long HDPE pipe mounted on the seafloor with an intake depth of 674 m. Remarkably, this CWP is still in operation today more than 30 years after its deployment in 1987. The design of the NPPE plant itself was a collaborative effort between two DOE laboratories, the *Solar Energy Research Institute* (SERI) and *Argonne National Laboratory* (ANL), and a Honolulu-based non-profit company, the *Pacific International Center for High-Technology Research* (PICHTR). Figure 7 is an aerial view of the land-based OC-OTEC plant as built, which operated for five years from 1993 through 1998; Fig. 6 represents a sample of the data collected. Grid-connected gross power production peaked at 255 kW, with a net power production estimated at just over 100 kW, due to unavoidable built-in inefficiencies in the seawater distribution layout. These figures still represent OTEC records to date. A seawater desalination module also was tested. NPPE equipment generally performed according to specifications, except for the vacuum pumps which required major design changes and adjustments.

As the price of oil stabilized and political priorities in the US changed after 1980, the effort to develop OTEC into commercial reality essentially waned by the mid-1990s. The responsibility for further OTEC development implicitly was left to the private sector. Hence, the ambitious programs that aimed to establish the technical viability of pre-commercial OTEC systems have not yet been satisfactorily completed. Floating OTEC pilots of 5 to 10 MW that would be operated for a few years still need to be constructed. Otherwise, it is unlikely that a private company would assume the financial risk of investing in commercial OTEC plants while the technology has an insufficient track record. Figure 10 shows, however, that from an economy-of-scale point of view, pilots remain unfavorably situated. A lack of commercial maturity and large investment needs (at about $300 million for a significant pilot) clearly suggest that only a public effort is likely to move OTEC technologies forward. This will require a strong political will. However, the geographic location of OTEC resources may represent a serious handicap from the perspective of most wealthy technologically developed nations. Following a sharp rise in the prices of fossil fuels through the first decade of the 21st century, several industrial and governmental stakeholders have recently been trying to realize the objective of deploying OTEC pilot plants, but their efforts have yet to proceed beyond the design stage.

References

1. J.-A. d'Arsonval (1881). Utilisation des forces naturelles — Avenir de l'électricité. *Revue Scientifique*, **17**, 370–372 (in French).
2. G. C. Nihous (2010). Mapping available Ocean Thermal Energy Conversion resources around the main Hawaiian Islands with state-of-the-art tools. *Journal of Renewable and Sustainable Energy*, **2**, 043104, 1–9. http://jrse.aip.org/jrsebh/v2/i4/p043104_s1?view=fulltext.

3. http://www.globalwarmingart.com/images/b/b0/Thermohaline_circulation
   .png.
4. A. De Vos (1984). Efficiency of some heat engines at maximum-power conditions. *American Journal of Physics*, **53**(6), 570–573.
5. G. C. Nihous and L. A. Vega (1996). Performance Test Report: Analysis of representative time history records obtained at the USDOE 210 kW
   OC-OTEC Experimental Facility in the power production mode. U.S.
   Department of Energy Report, DE-AC36-92CH10539, 81.
6. G. Claude (1930). Power from the tropical seas. *Mechanical Engineering*,
   52(12), 1039–1044.
7. L. R. Berger and J. A. Berger (1986). Countermeasures to microbiofouling in
   simulated Ocean Thermal Energy Conversion Heat Exchangers with surface
   and Deep Ocean Waters in Hawaii. *Applied and Environmental Microbiology*,
   51(6), 1186–1198.
8. L. A. Vega (1995). The 210 kW Open-Cycle OTEC Experimental Apparatus:
   Status Report., *Proceedings Oceans '95 Conference*, San Diego, 6.
9. J. Y. Cousteau and H. Jacquier (1981). Énergie des mers: Plan-plan les
   watts. Chapter 9 in *Français, on a volé ta mer*, R. Laffont publisher, Paris
   (in French).
10. G. C. Nihous (2007). A preliminary assessment of Ocean Thermal Energy
    Conversion (OTEC) resources. *Journal of Energy Resources Technology*,
    129(1), 10–17.
11. K. Rajagopalan and G. C. Nihous (2013). Estimates of global Ocean Thermal
    Energy Conversion (OTEC) resources using an Ocean General Circulation
    Model. *Renewable Energy*, **50**, 532–540.
12. K. Rajagopalan and G. C. Nihous (2013). An assessment of global Ocean
    Thermal Energy Conversion (OTEC) resources with a high-resolution Ocean
    General Circulation Model. *Journal of Energy Resources Technology*, **135**,
    041202, 9.
13. K. Rajagopalan and G. Nihous (2013). An assessment of global Ocean
    Thermal Energy Conversion resources under broad geographical constraints.
    Journal of Renewable and Sustainable Energy, **5**, 063124, 11.
14. Y. Jia, G. C. Nihous, and K. Rajagopalan (2018). An evaluation of the
    large-scale implementation of Ocean Thermal Energy Conversion (OTEC)
    using an ocean general circulation model with low-complexity atmo-
    spheric feedback effects. *Journal of Marine Science and Engineering*, **6**, 28,
    doi:10.3390/jmse6010012.
15. R. A. Locarnini, A. V. Mishonov, J. I. Antonov, T. P. Boyer, and H. E. Garcia
    (2006). World Ocean Atlas 2005, Volume 1: Temperature. NOAA Atlas NES-
    DIS 61, S. Levitus editor, U.S. Government Printing Office, Washington, 182.
16. Wikipedia. http://en.wikipedia.org/wiki/North_Sea_oil.
17. L. A. Vega (2007). The economics of Ocean Thermal Energy Conversion. *4th
    EnergyOcean Conference*, Turtle Bay Resort, Oahu, Hawaii.
18. H. A. Avery and C. Wu (1994). Renewable energy from the Ocean — A Guide
    to OTEC. Oxford University Press, New York, 446.
19. G. Claude (1957). Ma vie et mes inventions. Plon, Paris, 272 (in French).

20. R. Baillot (2010). Georges Claude: le génie fourvoyé. EDP Sciences et Histoires, ISBN 978-2-7598-0396-5, 490 (in French).

21. W.E. Pittman, Jr. (1982). Energy from the Oceans: Georges Claude's magnificent failure. *Environmental Review*, **6**(1), 2–13.

22. B. Macdonald (1981). Ocean Energy Launch 1980, a two-part YouTube movie: http://www.youtube.com/watch?v=8aQXg5M5DiM and http://www.youtube.com/watch?v=9yRWNQ4OJDo.

23. T. Bjelkeman-Petterson (2007). *OTEC in Nauru*, a two-part YouTube movie: http://www.youtube.com/watch?v=_mGOcqofERM and http://www.youtube.com/watch?v=HWVWD80ENdM.

Chapter 7

Wave Energy

Arne Vögler

Hebrides Marine Services Ltd
Lews Castle College — University of the Highlands and Islands
Stornoway, Isle of Lewis, Outer Hebrides, Scotland, UK HS2 0XR
arne.vogler@btinternet.com

Abstract

In recent years the global endeavor to exploit the vast marine energy resources from waves and tides has seen a considerable number of very promising developments. This chapter provides an overview into the wider context of the wave energy conversion (WEC) sector, and an introduction into some generic WEC device-specific characteristics is given. In addition to the prime mover technology, i.e. the part of a WEC that extracts the impacting energy from the waves, examples are also given for widely used power take-off systems, which are required to convert the wave energy to useful forms of power. Examples of recent and ongoing sectoral developments across the world are described, and introductions into the wave resource assessment process and considerations for wave power project development are also provided.

1 Introduction

Water waves of interest for energy production are created by the interaction of wind with the sea surface. A wind that blows over the sea initially results in small ripples known as capillary waves with a wave period of less than 0.1 seconds. Wave period describes the time required by a wave to pass a fixed point. Prolonged exposure of capillary waves to a wind results in an increase of the period and, depending on the wind speed, also a gradually growing wave height. Wave height refers to the distance between the lowest and highest parts of the wave (wave trough and wave crest). Waves with a period of between 1–30 seconds are called ordinary gravity waves and a distinction can be made between wind waves and swell waves. The former term describes locally generated waves, where the wind continues to interact with the sea, and these are typically of a shorter period than swell waves. Although virtually no mass is transported by water waves, the waves travel at considerable velocities, which can easily exceed 15 m/s in case of ocean swell waves. Therefore it is often seen that waves progress out of the area in which they are generated to travel long distances. As the waves move along, the wave pattern is purified and a transition from a somewhat more chaotic sea surface in the area exposed to the wind to a more regular wave pattern is observed. Energy in the traveling waves is largely conserved, and only a modest decrease of wave height is seen with distance from the origin. Swell waves can travel thousands of kilometers, and it is not unusual to encounter waves in northwest Europe that originate from low pressure wind systems near Newfoundland. Similar observations can also be made on the west coasts of other continents, and it is these long traveled swell waves that are of greatest potential for the conversion of wave power to electricity or other uses.

With 71% of the Earth covered by oceans and a large amount of this area exposed to considerable wind regimes and resulting high waves, the global theoretical wave power resource has been estimated[1] as 3.7 TW. Based on this theoretical resource, and compared against the worldwide electricity supply figures from 2008, wave energy absorbed by the world's coastlines is almost twice the electrical energy demand.[2] Or in other terms, the global wave resource is higher than the output of 1,000 modern nuclear power stations (e.g. based on the 3.2 GWe rated UK Hinkley Point C project). In reality however, it will not be possible to exploit the full wave resource around the world, and conversion losses from input wave power to grid quality electrical supply also present a further reduction to the effectively useful wave resource. But there should be little doubt that following the successful development of the wave energy conversion technology, this resource can be a major contributor to the world's energy requirements.

This chapter gives an introduction to the wider context of wave power developments, including details on wave resource characterization, Wave Energy Converter (WEC) technology, development challenges and outlook.

2   Political and Societal Context

The combination of an increasing global population with overall economic growth in the world results in a growing demand for energy. This is required to satisfy both domestic and commercial uses such as heating, manufacturing, road, rail and air transportation, or shipping. To date, by far the biggest amount of this energy is derived from finite fossil resources, including oil, gas and coal. Although these finite resources are still sufficient to supply global energy needs for a good number of years, eventually an alternative source of energy is required to maintain security of the energy supply. The use of fossil fuels has shown numerous negative side effects, primarily related to pollution, and it is generally accepted that the burning of fossil fuel contributes to global warming through the release of greenhouse gases.

To alleviate the threat posed by the increased release of greenhouse gases, the Kyoto protocol, an international agreement ratified by most developed nations in 1997, came into force in 2005. The aim of this agreement was to reduce global greenhouse gas emissions with individual target levels agreed to by the signatories. For example, based on 1990 levels, the European Union and the USA committed to reduce emissions by 8% and 7%, respectively, by 2012. However, the USA retreated from the agreement in 2001 to develop an individual strategy on emissions reduction. In 2015, the Kyoto protocol was superseded by the Paris Agreement, when 195 countries adopted a legally binding climate deal including a commitment for further reductions of emissions. Although it is appreciated in the agreement that some developing countries are likely to see an increase of greenhouse gas emissions in the short term, this is to some extent mitigated by considerable emissions reduction commitments by others. For example, Scotland has set emission reduction targets of 42% for 2020 followed by a further reduction of 90% of the initial threshold value by 2050. The urgent requirement to reduce carbon-based and other greenhouse gas emissions to reduce the level of global warming was further reinforced in a report by the UN Intergovernmental Panel on Climate Change (IPCC) in 2018.

Low carbon solutions for the generation of electricity include nuclear, hydro from dammed or natural reservoirs, geothermal, solar and marine energy from waves and tides. Following a major incident at the Fukushima Daiichi nuclear reactors in Japan in 2011, the global effort to pursue nuclear power has been reduced, which has led to an increased interest to develop renewable energy solutions. The installed capacity for mature technologies such as wind generators, solar and geothermal systems has considerably increased over recent years, in parallel with a strong research effort to achieve commercial market readiness of other forms of renewable electricity generation. The globally installed marine energy generation capacity has more than doubled from 12 MW in 2016 to 25 MW in 2017.[3] An industrial goal has been set by the executive committee of Ocean Energy Systems to have an installed worldwide marine energy capacity of 300 GW by 2050. This is anticipated to directly support 680,000 jobs and lead to an annual reduction of 500 million tonnes of CO_2 emissions.[4]

Due to the very limited visual presence of marine energy developments, these are generally perceived more positively by the public than large-scale wind farms. Although some concerns are being raised in relation to a possible environmental impact, this is a topic still under review and evaluation, and to date limited evidence has been produced to demonstrate such impact. Risks might be related to possible collisions of wildlife with generators, although this seems to be of higher relevance for in-stream tidal energy developments than for wave energy extraction. Other possible side effects of marine energy developments are seen in the displacement of fishing activities at development sites. This is sometimes counter-argued by suggesting that fisheries' exclusion zones around marine energy sites can support the recovery of dwindling fish stocks by providing a safe habitat. Concerns are also sometimes raised with a view to the impact of marine energy developments on navigational safety or, in the case of wave power extraction, on reduced wave heights down-wave of developments, which is of concern to some members of the surfing community. At the same time, such a wave height reduction has been shown in numerical models to alleviate the threat of coastal erosion, which is of great relevance to many coastal communities. Overall, the anticipated increased economic activity in remote coastal regions, as a result of marine energy developments, is looked at very favorably by communities and policymakers. The economic upswing of the Scottish Orkney Islands following the launch of wave and tidal energy test sites at the European Marine Energy Centre (EMEC) in 2003 is an excellent example of how rural areas can greatly benefit from the development of marine power solutions.

Generally, there appears to be widespread agreement that wave and tidal energy sources should feature strongly in future energy supply scenarios, and the potentially achievable benefits seem to largely outweigh any possible drawbacks of marine energy installations. The biggest challenge remains the identification of cost-effective marine energy technology to achieve market competitive solutions that allow the investment case to be made for large-scale project development.

A large number of government initiatives around the world are actively promoting the development of the wave energy sector. One of the earlier programs was the Scottish Government-backed Saltire Prize, launched in 2008 and offering a £ 10Million (US$14 m) award for the first company to achieve an output of 100 GWh of electricity over a continuous 2-year period in Scottish waters before 2017. However, by 2015, and following the cessation of WEC development activity by the then leading wave power companies Pelamis, Aquamarine and Voith Hydro Wavegen, it became clear that no company would achieve the result. The Saltire Prize has subsequently been replaced by Wave Energy Scotland (WES), a government-funded body to support wave power-related research and development activities. Until 2018, WES had invested £ 39 m (US$ 54 m) over 86 contracts to 177 separate organizations across 13 different EU countries. The aim of this program is to identify and help develop the most cost-efficient solutions for entire WEC systems, including on subcomponent levels, such as power take-offs (PTO), moorings or materials. A similar approach has been taken by the US Department of Energy (DoE) when launching the Wave Energy Prize as a competitive program. Following a shortlisting of nine out of 92 initially applying companies, finalists in the program received funding and support to demonstrate their technologies in laboratory-scale model tests. The most successful of the applicants was awarded a cash prize of $1.5 m, followed by awards of $500 k and $250 k for the second and third prize. Other programs to support technology development, wave resource characterization and project development are available in many other coastal states, including in North and South America, Europe, Australia, Asia and Africa.

3 The Resource

The efficient conversion of wave power to useful energy requires a strong understanding of the local wave resource at a project site. Depending on where wave power farms are planned for construction, not only has the annual average wave power distribution to be considered, but attention must also be given to seasonal variation, extreme wave heights, short to medium-term predictability of the resource and the availability of suitable infrastructure. Although seasonal variation of wave heights might be of less concern in areas such as the trade wind regions, very significant differences in the wave climate can be observed between summer and winter towards the higher latitudes, such as in the northern parts of Europe or North America.

To quantify the wave power potential at a given site, the distribution of individual waves over both relatively short time periods and the long term variation have to be considered. The first is typically done over half hourly periods, and referred to as a sea state. The most basic key parameters used to describe sea states are the significant wave height and wave zero-crossing period. Traditionally, the significant wave height was defined as the average of the highest third of waves as seen by a trained observer. Following the introduction of technology to accurately measure wave time series data, e.g. by pressure gauges, staff poles or floating accelerometer-based sensors, the same methodology is still in use, albeit with the trained observer being replaced with sensor data. However, other methods to calculate the significant wave height offer valid alternatives. These include spectral analysis of the sea state, or a statistical approach, where the significant wave height is established as a multiple of four of the standard deviation of a sea surface elevation time series of a sufficient sample rate.

An example of a wave time series from a Datawell Waverider buoy is shown in Fig. 1. For clarity, only a short three-minute subset of a half-hourly record is shown. These data were obtained from a set of accelerometers along the vertical, northerly and westerly axes inside the buoy, and displacement values are calculated by a process of double integration of the measured acceleration values.

Another more complex method to calculate the significant wave height and other parameters of a sea state considers the frequency distribution of individual waves over an observation period, known as a wave spectrum.

By processing time series data as shown in Fig. 1 based on a Fast Fourier Transform (FFT), the directional wave power density across the frequency spectrum can be obtained. An example of this is given in Fig. 2, where the gray shaded area indicates the energy across the dominant frequencies in the sea state, and the dashed line, the direction of the approaching waves at the given frequency. An understanding of the directional distribution of waves is important to wave power extraction, as it can often

Roag_wavegen, 2011-12-08 17:30:00

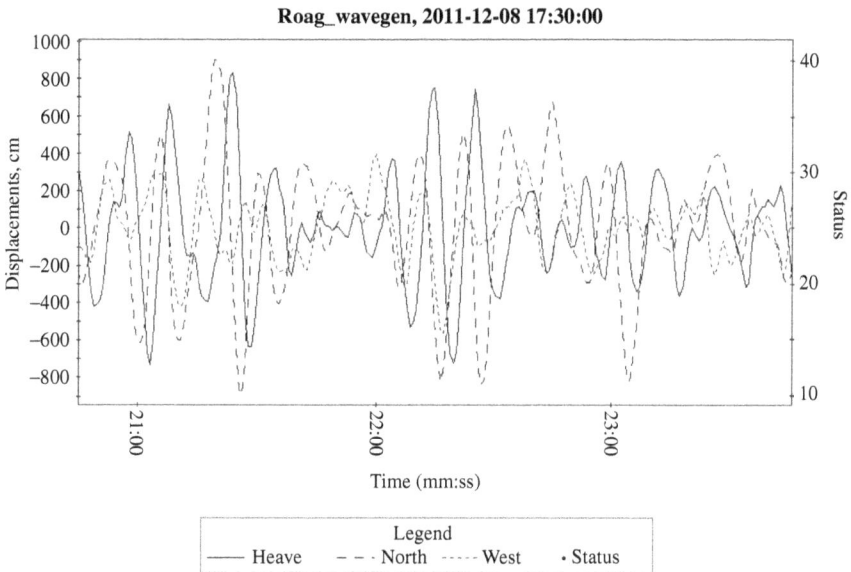

Fig. 1.   Example for a wave data time series. Shown are the vertical (heave), northerly and westerly displacement values of a floating sensor over three minutes.

Directional wave spectrum

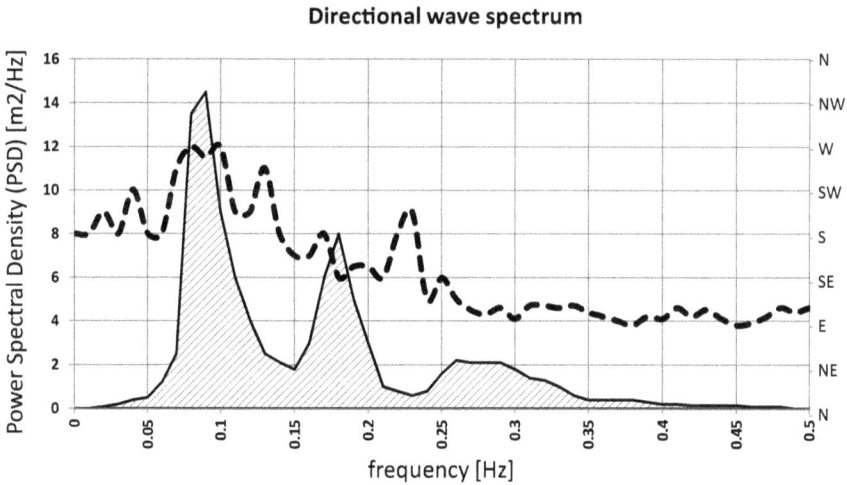

Fig. 2.   Spectral distribution of a sea state following an FFT-based process.

be observed that high energy swell waves approach a potential energy site from a different direction than locally generated wind waves. Depending on the wave energy converter solution proposed for a site, such a bimodal sea state can have a considerable impact on energy generation and must be carefully considered.

Provided that spectral data such as shown in Fig. 2 are available, key sea-state parameters can be calculated from the individual moments of the underlying data. These moments are calculated as shown in Eq. (1), where M_n represents the spectral nth moment, n the order of moment, f_j the individual frequency bins, Δf_j the frequency spacing between bins and $PSD(f_j)$ is the Power Spectral Density at the individual frequency bins.

$$M_n = \sum_j PSD(f_j) f_j^n \Delta f_j \qquad (1)$$

Once the moments are established from the wave spectrum, the spectrally derived significant wave height can be calculated as per Eq. (2).

$$H = 4\sqrt{M_0} \qquad (2)$$

All of the approaches described above return reasonably similar values and can generally be interchanged, depending on the requirement for data analysis and data availability. In the context of wave energy, preference is generally given to the spectrally derived parameters, and this is also widely reflected in the definitions and parameterizations used in numerical software solutions for wave analysis.

Similar to the wave height, traditionally the wave zero-crossing period was also established based on a statistical analysis of observed or measured wave time series data. This parameter describes the average duration over which waves in a time series record cross the mean water level in the same direction. A difference is sometimes made between zero up-crossing and zero down-crossing periods. These distinguish between the time taken from the moment when a wave crosses the mean water level upwards, until the next upwards crossing occurs, or for the down-crossing period, the average time lapse between subsequent downwards crossings. If wave spectral data are available, the zero-crossing period can also be calculated from the power spectral density across the frequency spectrum as shown in Eq. (3). Other definitions for wave period include the peak period, i.e. the period of waves with the highest energy content in a sea state, or the widely used energy period, which describes the period of an equivalent monochromatic wave that features the same energy as the observed irregular sea state.

$$T_z = \sqrt{\frac{M_0}{M_2}} \tag{3}$$

With knowledge of both wave height (H) and period (T) of a sea state, the wave power in deep water can be calculated as shown in Eq. (4).

$$P = \frac{\rho g^2 T H^2}{64\pi} \tag{4}$$

where ρ represents the density of sea water (typically taken as $1,025 \text{ kgm}^{-3}$) and g the gravitational constant of 9.81 ms^{-2}.

Waves start to interact with the seabed when the wavelength is equal or smaller than twice the water depth, and in those situations equations adequate for deep water conditions are no longer valid. For ease of access and to avoid high costs associated with long electrical subsea cabling, wave power developments are mainly planned in coastal waters, relatively close to shore. Although the simplified deep water wave equations can still be used in those situations for an initial assessment, the use of a modified and more complex wave power equation than shown above yields more accurate results. This is shown in Eq. (5), which is valid for all but very shallow water depths.

$$P = \rho g \frac{H^2}{16} C_g \tag{5}$$

To accommodate for interactions between waves and the seabed during shoaling conditions, the wave group velocity C_g is now incorporated in the calculation, as follows:

$$C_g = 0.5 \left(1 + \frac{2kd}{\sinh(2kd)} \right) c \tag{6}$$

where k is the wave number, d the water depth, *sinh* the hyperbolic sine function and c the individual wave velocity. To work out k, knowledge of the wavelength λ is required. Equations (7) to (9) show how to calculate values for k, c and λ, respectively. It is worth noting that the wavelength and wave velocities are fully independent of the wave height, and are predominantly defined by wave period. It is further highlighted that an iterative solution is required to establish the wavelength.

$$k = \frac{2\pi}{\lambda} \tag{7}$$

$$c = \frac{gT}{2\pi} \tan h \left(\frac{2\pi d}{\lambda} \right) \tag{8}$$

$$\lambda = \frac{gT^2}{2\pi} \tan h \left(\frac{2\pi d}{\lambda} \right) \tag{9}$$

The previous section has shown how to work out relevant sea-state parameters, based on statistical or spectral data. Such data when obtained from field observations are mainly limited to single point measurements from in-situ sensors, and thus do not reveal information about the spatial distribution of wave power across a potential project site. Although some wave sensors are capable of providing a certain extent of spatial coverage, such as X-band or HF radar technology, these are often not as accurate as spot sensors, and provide only a relatively coarse resolution. Waves can also be measured with satellite-based systems, but as satellites constantly move around the Earth's orbit, it is not possible to obtain continuous data for a fixed location.

To support wave power developers with the production of comparable yield forecasts, an internationally agreed standard on wave resource assessment has been produced under the auspices of the International Electrotechnical Commission (IEC).[5] A standardized approach is important to not only provide comparability between sites, but also to improve the credibility of return on investment (RoI) predictions, an important aspect to secure project finance for developments. Standard IEC/TS 62600-101 clearly defines specific requirements for different classes of resource assessments, which are dependent on the planning and implementation stage of planned developments. Resource assessments are distinguished between reconnaissance, feasibility and design stages, with areas under investigation reducing in size from 100s of kilometers during reconnaissance to specific project sites of less than 25 km extent in the design phase. As it is impractical and not cost-effective to obtain high resolution wave data for a large area with in-situ sensors, a range of computer models have been developed to produce maps showing the distribution of required wave parameters across both spatial and time domains. Numerical modeling platforms for wave resource assessment, such as the widely used Delft 3D, SWAN or DHI Mike 21 SW, provide solutions by applying a phase-averaged approach over the

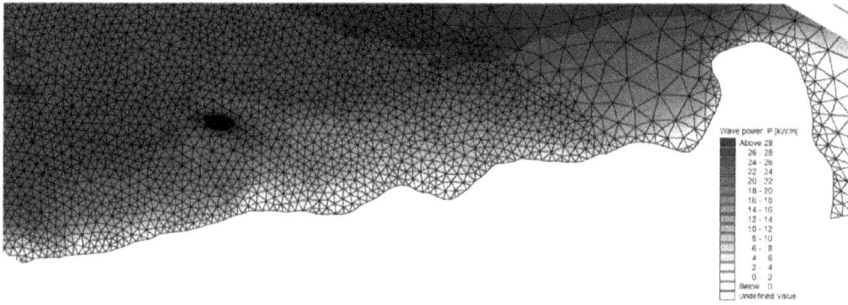

Fig. 3. Wave power distribution at Isle of Lewis, Scotland, on Mike 21 SW software.

model domain. The area of interest is discretized into a mesh with individual mesh cells of variable size to allow for sufficient detail at the core areas of interest, while reducing the computational time by utilizing a larger mesh in the surrounding areas. An example is shown in Fig. 3, where the project target site is based in the area covered with the smaller triangular mesh. Phase-averaged models, also referred to as spectral wave models, calculate the progressing waves across the mesh from the outer model boundaries by using the source term Eq. (10). The required parameters are calculated for each mesh cell, where S represents the energy source term, S_{bou} the energy input from the nearest boundary, S_{Wind} the energy input from wind–water surface boundary interaction, S_{nl3} and S_{nl4} show the energy transfer from triad and quadruplet wave-to-wave interactions, respectively, S_{ds} and S_{bot} represent energy dissipation through white-capping and bottom friction, and S_{surf} is used to describe energy losses through shoaling-related wave breaking.

In situations, where detailed wave information is required across a large region, and over prolonged periods of time, computer-based numerical models offer a cost-effective and reliable alternative to field measurements.

$$S = S_{bou} + S_{Wind} + S_{nl3} + S_{nl4} + S_{ds} + S_{bot} + S_{surf} \tag{10}$$

Following the global ambition to exploit the vast wave energy resource, a large number of numerical wave models have been produced providing coverage of the global wave resource in a coarse approach up to high resolution localized site-specific model domains. An indicative example of the global annual averaged wave power resource based on model data is shown in Fig. 4. It can be seen that the areas with the highest energy availability are in the southern oceans, where the wave fetch can largely develop unhindered

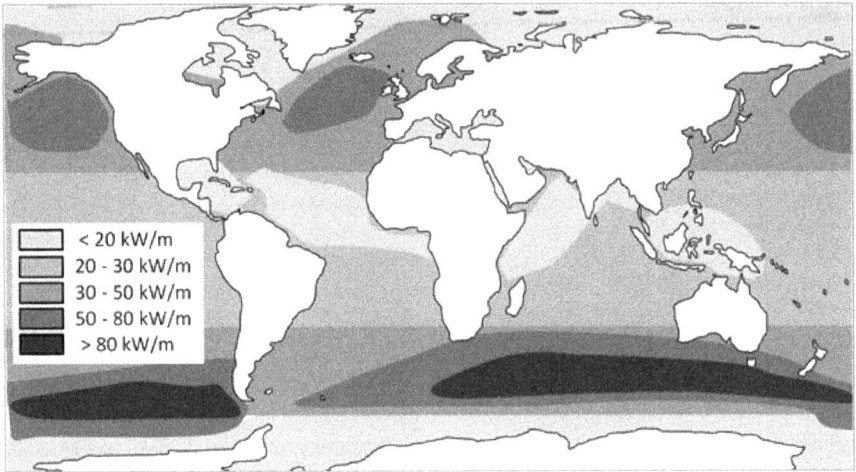

Fig. 4.   World map of the annual averaged wave power resource.

by continental landmasses. But as access to shore and population centers is essential for successful project developments, the most promising sites for commercial wave power extraction are in the northwestern regions of Europe and America, where the available resource is closer to areas with a relatively large population density. Other promising sites that match available wave power with local demand are found in Chile, South Africa and the south of Australia. But as these areas with the highest wave power potential are exposed to a strong seasonal variability, with destructive storms in winter and more benign sea states during summer, deployment of a technology that works with a high efficiency all year round is difficult. This is different for the areas closer to the equator, where a more modest resource of a relatively annual constant nature potentially offers considerable advantages to project developments.

4   WEC Technology

The exploitation of wave energy to provide solutions for a range of challenges has been of interest for a long time, with an estimated 340 patents granted for wave-powered generators between 1856 and 1973.[6] Although no specific technology has so far achieved full competitive market readiness with a view to generation of electricity, wave power has been successfully used in other applications.

Examples of early devices that utilized passing waves for a useful purpose include Brown's Bell Buoy, which was invented around 1850 as a navigational aid. In this simple device, a bell was attached on a pole to a floating buoy to create an audible warning to shipping as the system rolled in the waves. This simple method was enhanced some years later by US Admiral Cortenay, who utilized a separate hollow chamber in the buoy, which was open at the bottom, and had a small opening above the water line. As the waves pass the buoy, the water level inside the chamber rises and falls, thus creating an oscillating airflow at the opening above the surface. By placing a whistle or fog horn inside the airflow, a loud sound is generated to safeguard shipping in hours of darkness or poor visibility. Both concepts are still being used today for the same purpose, with the latter approach also being explored for the generation of electricity.

4.1 *Power extraction*

When waves encounter an object, the impacting wave energy is partially transmitted, reflected (or deflected) or absorbed. In the context of wave energy conversion, the aim is to maximize the percentage of absorption, as this equals the energy available for the conversion to electricity or other beneficial uses. Wave reflection from WECs is generally not desirable, as interference with the incoming waves might confuse the impacting sea state and lead to a reduction of the useful components of the available resource. The partial transmission of waves by a WEC is not considered a major setback, as subsequent WECs in an array layout can be deployed to further absorb the passing wave components. Besides other considerations, such as survivability, or costs for manufacturing, installation, operations and infrastructure, WEC design is strongly driven by the goal of achieving a high extraction rate, while causing minimum wave reflection.

To maximize the energy extraction ratio from waves, it is important to match the WEC natural frequency against the dominant wave frequency. As the wave frequency is not static and varies over time, an ability to adjust the natural frequency of a WEC to retain system resonance is highly desirable. The natural frequency of a system is given by Eq. (11), where k is the stiffness as defined in Eq. (12) and m is the mass of the system. Force and displacement are represented in Eq. (12) by F and δ, respectively.

$$f = \frac{1}{2\pi}\sqrt{\frac{k}{m}} \tag{11}$$

$$k = \frac{F}{\delta} \tag{12}$$

From Eq (11) it can be seen that to adapt the natural frequency, a change of either the stiffness or mass is required. Although it is possible to dynamically adjust the mass of a WEC by ballasting the system with water, this is only practical for medium-term adaption to general sea states. To achieve a fast response, as is required for system tuning on an individual wave-by-wave basis, dynamic changes to the stiffness are required. This can be achieved through adjustment of the power take-off (PTO), either by flow control in a hydraulic system, tuneable damping on the electrical conversion train, or in the case of wave-powered water desalination plants, by adjustment of the active area of a membrane filtration system. A wide range of WEC control strategies have been proposed and tested over recent years, and this is still an active research area. Approaches range from a very simple static solution, where the WEC is tuned to annual average parameters, without the ability to change during operations, up to the use of fast adaptive control functions that are based on short-term forecasting of impacting individual waves to always ensure an optimized response. The further development and implementation of efficient and practical control system solutions can be seen as an integral part of the process of the commercialization of the WEC technology.

4.2 *Project site considerations*

WECs can be categorized in a number of different ways. From a project developer's view, the first assessment will likely be driven by the location of the installation site. Of the wide range of conceptual and already existing technologies, some are most suitable for shore-mounted installations, e.g. integrated into a breakwater at a harbor, or a cliff-faced coastline. Other devices work at their best efficiency in shallow water installations in the nearshore environment, and another group of WECs requires deeper water at a larger distance from the coast.

Shore-mounted devices offer the advantage of ease of access, with parts of the installation being installed well above the water line on solid ground, thus reducing the risk of failure due to exposure to sea water. An example of a breakwater integrated WEC solution is the world's first commercial wave power plant in Mutriku (Basque Country, Spain), which was designed and supplied by Voith Hydro Wavegen and commenced operations in 2011.[7] The 300 kW rated system is based on the oscillating water column (OWC) concept, where the water level in 16 hollow chambers rises and falls with the wave action, thus creating air pressure fluctuations inside the chamber.

The wave-pressurized air is passed through a Wells turbine, a turbine that rotates in the same direction regardless of direction of airflow, thus generating electricity.

A WEC concept typically designed for the nearshore shallow water environment is the oscillating wave surge converter (OWSC). By mounting a flap perpendicularly aligned to the incoming waves on the seabed, the increased horizontal surge motion of waves during the shoaling process pushes the flap back and forth. Power is generated by the motion of the oscillating flap through a piston hinged between the moving flap and fixed WEC structure. An example of an OWSC is the 800 kW Oyster device by Aquamarine Power, which was installed at the European Marine Energy Centre (EMEC) in Orkney, Scotland, in 2012.

Shared advantages of both near shore and shore-based installations are in the somewhat more aligned wave climate, compared to offshore sites. Modification of waves occurs during the shoaling process due to diffraction and refraction, i.e. the interaction with the seabed and nearby headlands or reefs. This generally results in a relatively uniform direction of the waves closer to shore, and thus allows the installation of fixed WEC systems that are unable to adjust to directional changes of the incoming wave front. WEC installations in shallower water close to shore feature less exposure to extreme wave events, which is a further distinct advantage of such locations. Depending on the wave steepness, i.e. the ratio between wavelength and height, wave breaking occurs when waves become too steep and thus unstable. As waves undergo the shoaling process, both wavelength and height are reduced through bottom interaction. This results in an increased steepness, with larger waves dissipating part of their energy when progressing from the offshore to the nearshore zone. Although this results in a less energetic wave climate at WEC project sites close to coast, a narrower distribution of wave heights can be observed, with maximum wave heights limited by the water depth. This narrower wave height distribution generally allows for a more efficient WEC tuning to produce a higher yield. In addition, the avoidance of extreme wave events is an important consideration to WEC survivability. On the downside of nearshore installations, it cannot be ignored that less energy is available, and a potentially higher overall yield may be achieved further offshore.

WEC systems targeted for installation in deep water offshore locations require the ability to operate efficiently under varying wave directions. This can either be achieved as an inherent design feature, such as in an omnidirectional point absorber WEC (e.g. OPT's Power Buoy), or by the ability to directionally realign the WEC to match the incoming wave direction. Although the full exposure to waves that have not experienced energy losses during the shoaling process potentially allows for a higher power generation compared to down-wave sites near shore, this is offset by a potentially reduced overall performance due to the wider range of encountered sea states. Careful consideration is required to test how offshore WECs can handle extreme storm events, and an increased cost is generally associated with longer subsea cable routes and vessel transit time for construction, operations and maintenance.

4.3   *Prime movers*

An entire WEC system consists of a number of sub-components, including a prime mover device to convert wave power to mechanical power, and a power take-off system to convert the mechanical power to electricity or other uses. During the intense research efforts in recent decades, a number of different WEC prime mover concepts have evolved. Some of the more common types are described in what follows. As new technologies are still emerging under ongoing research and innovation programs, only a non-inclusive number of examples are shown.

4.3.1   *Submerged pressure differential*

These devices consist of a submerged floating body that is connected to the seabed by a tethered or fixed link. As the waves pass at the surface, the water pressure rises and falls, thus changing the buoyancy of the float. This change in buoyancy creates an upwards and downwards motion, which is converted into useful power by a hydraulic or electromechanical piston. As no parts protrude above the surface, these devices are well protected against breaking waves or extreme wave heights. Examples of submerged pressure differential WECs are the Australian Carnegie CETO device, or the Archimedes Wave Swing from AWS, Scotland.

4.3.2   *Point absorber*

Similar to submerged pressure differential devices, point absorbers also feature a floating body connected to the seabed. The surface-based buoy follows the wave motion, and a number of different concepts of how to convert this motion to useful energy exist. Where some point absorbers generate power through the wave-driven changing distance between float and seabed, others feature a horizontal disk or similar device suspended from the buoy in the water column. As the buoy is exposed to the waves, the suspended mass is constrained to follow this motion due to its vertical drag. This results in a changing distance between the floating and submerged body, which is converted to useful power by a vertically moving rod. Another concept for a point absorber is the combination of a long spar buoy linked to a second buoy at the surface end section. The spar buoy motion is constrained by inertia and drag through the shape of the subsea section. This creates a difference in displacement between the two bodies, resulting in the surface buoy moving up and down the spar. This movement is translated into power through mechanical or magnetic coupling. Examples of point absorbers are the PowerBuoy from the USA-based OPT, or the CorPower device from Sweden.

4.3.3 *Oscillating wave surge converter*

By hinging a buoyant flap that extends across the water column at the seabed, the surge motion of passing waves create a backwards and forwards motion of the flap. Power is produced by a piston connected between seabed and moving flap. The piston can either be used to power a hydraulic or electromechanical generator directly at the WEC, or alternatively a liquid can be pumped to the shore to drive a hydroelectric Pelton wheel-based turbine. Examples of this type of device are the Aquamarine Oyster, Scotland, or the WaveRoller from the Finland-based company AW-Energy.

4.3.4 *Line attenuator*

Line attenuators feature a number of floating cylindrical segments that are interconnected by hinged hydraulic or electromechanical pistons and aligned perpendicular to the incoming waves. In an optimized configuration, the length of each individual segment is half the wave length, thus creating maximum displacement at the pistons as the waves pass by. The first line absorber with a number of design iterations extensively tested at full scale at different sites was developed by Pelamis Wave Power, Scotland.

4.3.5 *Oscillating water column*

The oscillating water column concept utilizes the rise and fall of waves inside a hollow chamber with an opening at the bottom to create fluctuations in air pressure. This changing pressure results in an air flow through a turbine situated at an outlet of the chamber above the water level. Systems can be based inside breakwaters, at shore-mounted structures, or also further offshore in dedicated floating vessels. The first company to successfully install and commission a commercial breakwater installation of an oscillating water-column system was Voith Hydro Wavegen (Scotland). A near-commercial-scale floating device is currently prepared for testing in Hawaii by the Irish company Ocean Energy.

4.3.6   *Overtopping device*

Overtopping devices use the wave run-up at the sloped edge of a WEC to fill a reservoir. This results in an elevated water level inside the reservoir, which is discharged at the bottom through a low-head hydroelectrical turbine. Devices can be shore-mounted or fully floating, and an example is the Wavedragon from Denmark.

4.3.7   *Other examples*

In addition to the examples mentioned previously, a number of other different or hybrid concepts are also undergoing various stages of development. In the Penguin device by the company Welly Oy from Finland, an eccentric mass inside a floating body is set in rotation by the irregular motion of the waves. This concept has been very successfully tested at EMEC in Scotland over recent years, and the company has secured further funding for the installation of additional devices in an array layout. Devices from Albatern (Scotland) or Grey Island Energy Inc. (Canada) can best be described as line point absorbers, where a number of floating bodies are interconnected via long rods or lattice structures. Other devices include point pivoted buoyant systems that can be connected to breakwaters to create oscillating motions as waves are absorbed. A development of up to 5 MW of individual segments of such a system using a hydraulic energy conversion train is currently underway in Gibraltar by the Israeli company Eco Wave Power. Another system following the same concept, but based on a linear electromechanical generator, is currently being developed and tested in sea trials in the Scottish Orkney Islands by the Italian company UmbraGroup.

Examples of the device types described above are shown in Fig. 5, and in Fig. 6, which shows an example of a hinged buoy designed by the Italian company Seapower and which was suitable for breakwater installation during lab testing in Italy in 2016.

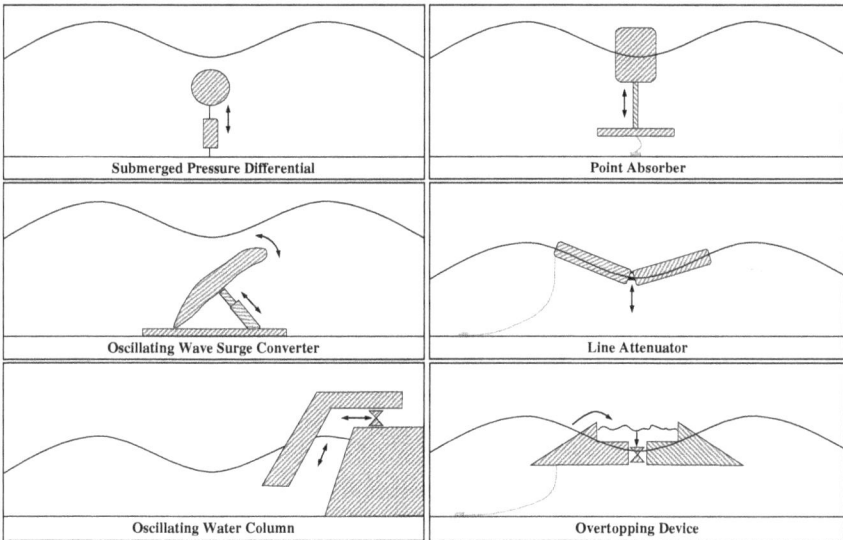

Fig. 5.   Examples of WEC prime mover categories.

Fig. 6.   Point pivoted WEC for breakwater installation.

4.4 *Power take-off*

Where early WEC devices mainly utilized hydraulic or pneumatic processes to convert wave power to useful energy, more recent developments have seen an increased use of electromechanical systems.

The direct conversion of mechanical power from the prime mover as suitable input into an electrical generator can be achieved by connecting a rotating eccentric mass to the shaft of a rotary electrical generator, or by the conversion from linear to rotating movement via a suitable gearbox arrangement. Another method of directly converting the linear power from a prime mover to electricity is through a ball screw generator, where the reciprocal movement of a wave-driven threaded shaft is used to rotate a ball nut as part of the generator assembly. Examples of these concepts that are currently undergoing successful test programs in sea trials are Welly Oy's Penguin, CorPower Ocean's heaving buoy and UmbraGroup's EMERGE device, respectively.

A conversion chain from wave power to electricity based on fluctuating air pressure was successfully demonstrated during an extensive test program by Wavegen. This was initially done at a dedicated oscillating water column chamber at a cliff face on the island of Islay in Scotland, and followed up with a commercial project in Basque Country, Spain. The conversion from wave to pneumatic pressure to generate electricity has the advantage of a generator assembly with only one moving part, which can be sited in a well-protected space above the water line. But to make such a system work efficiently, a turbine is required that moves unidirectionally, regardless of whether the air pressure inside the chamber rises or falls. A number of turbines have been developed to achieve this, with the Wells Turbine having been successfully installed and tested in a number of devices. A floating platform featuring a pneumatic power conversion chain connected to an OWC is currently being prepared for sea trials in Hawaii by the company Ocean Energy.

Hydraulic systems have been used successfully in a wide range of WECs during prototype testing, and have been the preferred option for a large number of technology developers. In a hydraulically based PTO, the waves are initially converted to mechanical power by the prime mover. This mechanical power is then used to drive a pump to circulate a liquid through a generator. Electrical generators that have been successfully used in this configuration include in-line systems, such as in the Pelamis device, or open systems, where the liquid is pumped to the shore to drive a

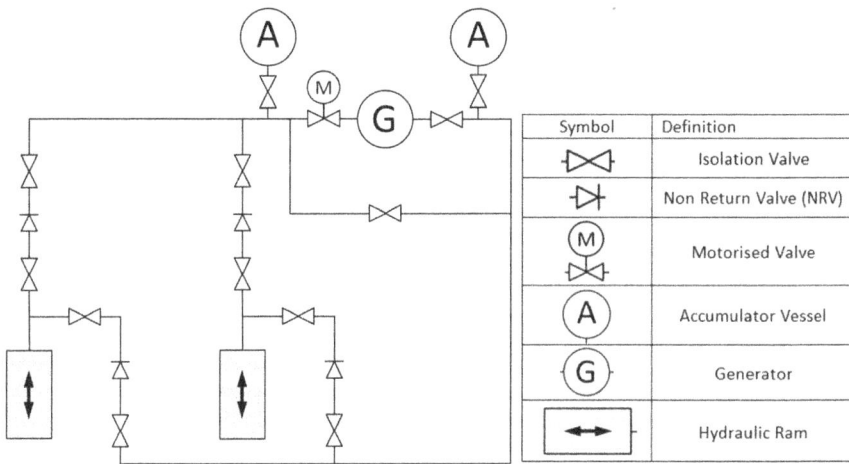

Fig. 7. Example of a basic hydraulic system for a WEC PTO.

Pelton Wheel or similar devices. Although hydraulic PTO systems include an additional conversion element, i.e. wave to mechanical to hydraulic to electrical power, compared to the more direct conversion from mechanical or pneumatic power to useful energy, good efficiencies have been reported. Similar to the challenge of an air turbine, it is preferable to achieve a uni-directional flow in a hydraulic PTO for WEC applications, and an example for a basic layout is shown in Fig. 7. By combining several hydraulic rams to one or more prime movers, and combined with accumulator vessels and non-return valves, a relatively steady positive pressure can be maintained to achieve a quasi-continuous output. Similar forms of power smoothing can be achieved in pneumatic or open flow hydraulic systems through the use of fly wheels, or in the case of electromechanical conversion systems, through short-term energy storage in the form of battery or capacitor banks.

Other PTO concepts consider the use of novel materials, including dielectric elastomers, and these are currently in the early stages of development. Wave power is also used for the production of potable water by pumping salt water through a membrane filtration system at high pressure, or by producing storable energy, e.g. hydrogen through the electrolysis of sea water. Wave power can also be used to pump water into high level storage reservoirs for release to a hydro-electric turbine when the demand for power arises. But these latter concepts all require an initial PTO as described earlier to either pump a liquid at pressure, or use electricity from electromechanical, pneumatic or hydraulic PTO systems.

5   State of the Art

At the time of writing this chapter (2019), the global effort to develop reliable and commercially viable wave power solutions is stronger than ever before. Although availability of private sector investment is limited, a large number of national or transnational public sector funded programs are actively supporting the research and development effort of specific technologies and dedicated test centers. International bodies such as the Ocean Energy Systems Technology Collaboration Programme (OES)[8] or Ocean Energy Europe (OEE)[9] promote collaboration and ensure adequate representation of the science community and industry stakeholders in governmental decision-making processes. Following on from open sea testing of WECs during individual projects, such as the Osprey device in 1995 or the shore mounted LIMPET installation in 2000 in Scotland, a dedicated test center for wave and tidal energy converters was opened in 2003 in the Orkney Islands. The European Marine Energy Centre (EMEC) features access to a number of test sites connected to the electrical grid, and fitted out with a range of instrumentation and data infrastructure. Due to the widely acknowledged success of EMEC, similar test centers have been implemented, and are now operable across the world, or are currently in advanced planning stages. Countries that currently feature operable dedicated test sites for ocean energy systems include the USA, Canada, Ireland, UK, Denmark, Australia and others. Additional test sites or individual WEC deployments in various planning or development stages exist in many countries with extensive coastal areas, including China, Sweden, Norway, Netherlands, Greece, Mexico, Korea and more. Due to the large number of national and international programs to support the development of wave energy technology, it is beyond the scope of this book to mention them all. Instead, an introduction to two examples of initiatives is given, namely the Wave Energy Prize of the US DoE and the WES support program in Scotland, UK.

The Wave Energy Prize competition opened for registration in 2015, and nine finalists were selected out of a total of 92 applicants. Aimed to attract new entrepreneurs into the wave power sector, as well as strengthening existing developers, the finalists were awarded access to a fully funded test program of 1/20th scale model devices. The goal was to compare and improve the efficiencies and reliabilities of a number of different devices, and in 2016 a US company AquaHarmonics was announced as the winner of the competition. The winning company was awarded a cash prize of $1.5 M, and following the success of the Wave Energy Prize competition, the company has secured additional funding of up to $5 M to work towards deployment of a full-scale device of their point absorber technology. Additional prizes of $0.5 M and $250,000 were awarded to CalWave Power Technologies and Waveswing America, the second and third winners of the competition.

The approach taken by the WES program differs from the US Wave Energy Prize by providing support not only to complete WEC systems, but also on a sub-component level. By awarding a series of research and development contracts through a competitive tender process open to any company from within the European Union, the aim of WES is to identify the most promising technology and subsystems that will lead to a breakthrough in commercialization of wave power. The program applies a stage-gate approach, where the awardees of successfully completed contracts are invited to apply for fully funded progression into a next, more advanced project stage.

In addition to the core technology work program, which supports novel WEC concepts, power take-off (PTO) systems, control systems and structural materials and manufacturing processes, WES has also commissioned a number of other specific studies. These are aimed at the identification and review of wave power-related issues, including electrical connections, moorings and foundations, or control requirements for WEC systems. The WES initiative was launched in 2014 and to date 200 organizations across 13 different countries have been involved in delivering contracts. Recent contract awards by WES include the further design, construction and half-scale testing of the Archimedes Waveswing, a fully submerged point absorber developed by AWS Ocean Energy, at EMEC in 2020. The Blue Horizon device, a floating hinged structure with an innovative hydrodynamic design to maximize energy capture, is scheduled for testing at the same time and under the same program by Mocean Energy. Under the WES funded PTO program, sea trials of a point absorber by CorPower Ocean AB have been successfully concluded in 2018. This project was focused on demonstration of the HiDrive PTO, which increases the WEC efficiency by an active phase control to ensure the point absorber maintains resonance with the incoming waves under normal conditions. At the same time, to protect the device in storm situations, the PTO is able to decouple the resonance between WEC and waves to prevent excessive movement and related high forces and stresses to the PTO, WEC structure and mooring arrangement. Sea trials of another innovative PTO under the WES program are to take place in early 2019. The EMERGE system by UmbraGroup uses a recirculating ball screw generator, incorporated into a point pivoted buoy prime mover for breakwater installations. This project has greatly benefitted from knowledge transfer from the aerospace sector, where systems similar to the EMERGE PTO are used in safety critical applications, e.g. as actuators to control wing flaps and rudders of planes. The aim of the WES program is to promote knowledge transfer by bringing together the most promising technologies under the program, including prime movers, PTOs or control systems, to finally achieve an optimized solution for cost-effective wave energy generation.

Other exciting developments and ambitions are being pursued in many other coastal states, e.g. the South Korean Ministry of Oceans and Energy is working towards an installed capacity of ocean energy technology of 1.5 GW, with a 220 MW contribution of wave power. An operable floating platform in China combines a 200 kW wave power system with solar energy input for the desalination of salt water.[10] Also in China, the Qingdao Pilot National Laboratory for Marine Science and Technology (QNLM) currently develops a wave and tidal energy test center with support from EMEC. Following the successful testing and demonstration of an array of submerged point absorber CETO devices by Carnegie Clean Energy in Australia, the company is currently developing for the installation of a next generation device at the Australian Wave Energy Research Centre (WERC) in Albany. The company's current plan is to follow up the next single device installation with a 20 MW wave farm, with consideration being given to a 100 MW expansion.[11]

The US company Ocean Power Technologies has recently installed one of their PowerBuoy point absorber devices in the Mediterranean Sea to provide a battery charging solution for Oil and Gas subsea operations. The same company has previously deployed and tested devices off the coast of the US and Japan, with the aim to provide a reliable power solution to offshore non-grid connected applications. A similar aim is being pursued with Albatern's WaveNET configuration of individual Squid devices. This system was designed to supply power to remote island communities or offshore aquaculture installations, which currently rely largely on on-board Diesel generators. An installed system of six 7.5 kW Squid devices is undergoing a long-term test and performance program in Scotland since 2016. Also, in an aquaculture context, the Newfoundland College of the North Atlantic (CNA) has investigated the use of a wave power-driven pump to supply sea water to shore-based fish cages. As aquaculture is a significant growth market globally, and with an increased awareness of environmental implications of large offshore fish farms, the development of onshore-based aquaculture systems is widely being considered. Such systems rely on a steady inflow of sea water, and this may well be provided through the utilization of wave power. Since 2018, the German company SINN Power has been operating a number of their breakwater-mounted point absorber devices in Greece, and is currently in the planning stages for another installation in the Caribbean. Also pursuing the combination of WECs with existing breakwater infrastructure is the Israeli company EcoWavePower, with an initial 10 kW research installation in Jaffa. Based on the successful completion of the Jaffa project, the company has secured a 5 MW connection agreement in Gibraltar, with the first development phase of 100 kW being operable since 2016. Other projects in Chile, China, Mexico and Scotland are also being pursued at varying stages of project development. These are just a few examples of wave power-related activity around the world and many other projects are also taking place. However, despite a large number of positive developments in the recent decade, the sector remains volatile, with a fast changing multitude of developers starting businesses, with varying levels of success.

6 Project Developments and Constraints

Once a WEC technology is ready to undergo testing or demonstration in a real sea environment, careful planning is required to identify a suitable deployment location. Although a growing number of dedicated test sites offer support with securing the legal deployment consents, electrical connection, monitoring and installation activities in more advanced stages for full commercial array deployment, it is generally up to the WEC project developer to carry out a comprehensive site assessment. Some of the key challenges in this context have already been described in previous sections, and these include the wave resource assessment, including the identification of localized energy hotspots. This chapter gives a brief introduction to additional considerations required for successful delivery of wave energy project developments.

6.1 *The investment case*

Probably the most important aspect of proceeding to the installation phase of a wave power development is the securing of project financing. Due to the political and societal frameworks that strongly support the shift away from carbon-based generation technologies to renewable forms of energy, a number of large multi-national utility companies agreed to invest significantly into wave power technology some years ago. At the time, world leading technology providers such as Pelamis Wave Power or Aquamarine Power successfully reached agreements with companies such as SSE, EON or Vattenfall to work towards commercial installation of their energy converters around 2010. Unfortunately, the ambitious expectations of delivering a reliable and efficient technology for large-scale installations could not be met in the given time frames, and as a result commitments to project finance were withdrawn. This shortage of funding was a key contributing factor for the closure of some of the leading companies, which was considered a major setback for the sector. Although public sector funding is still available on a competitive approach to support full-scale prototype development and testing, e.g. through the European Union H2020 program, this funding does not extend to full commercial WEC array installations. A successful approach taken recently in the tidal energy sector to raise finance is through the sale of unsecured bonds in the form of debentures. In this mechanism, individual investors agree to provide finance for a fixed period, with an attractive interest rate. However, the ability to successfully raise the required funding depends strongly on the credibility of the recipient, and recent developments in the wave power sector provide a positive outlook that debentures might also be an attractive tool to both project developers and investors in the near future.

The willingness of investors to engage in what is still seen as a high risk investment sector is driven by the anticipated return on investment in the short, medium and long term. A model widely used to evaluate the balance between investment and return is the levelized cost of energy (LCOE). This model compares the anticipated lifetime cost of a project with the amount of the electricity expected to be generated, and thus provides a useful tool for comparison between individual wave power projects and other forms of generation. Due to the pioneering nature of the wave power sector, no long-term and reliable figures for the LCOE exist to date. Recent studies from 2018 based on one-off WEC prototypes have suggested LCOE values for wave power in the range of $400 to above

$1,000 per MWh,[12, 13] and this compares with an LCOE for offshore wind energy developments of $240/MWh in 2018. The LCOE for offshore wind projects is anticipated to reduce to $120/MWh in the near future, and this can be seen as a target value for a commercially viable large-scale wave power sector. However, in the context of remote island locations, or other offshore power demands, where currently electricity is often generated from small to medium size Diesel generator systems at considerable cost, a higher LCOE for wave power projects might be competitive and acceptable.

6.2 Development consent

The approach to obtain consent for development of a wave power site differs between individual countries. Permissions are required from the owner of the seabed and the national regulator, which might be the same entity in some countries but not in others. In addition, consents are also required for any shore infrastructure, including landfall routes for cables and electrical connection points. Conflicting interests with fisheries or marine traffic have to be considered, as well as any potential environmental impacts or other stakeholder concerns. Consideration of the likely success of securing development consent for a potential wave energy project site is often the first step in project development, before a detailed resource assessment on a project scale is carried out. High level screening of areas of interest for development is generally based on the review of GIS (global information system) maps, which include various layers of constraints. Such constraints include designated environmentally protected areas, seabed composition, main shipping routes, unexploded ordnance or other types of spoil ground. Once a high level screening process has identified one or more suitable development sites, the detailed process of securing consent from the seabed owner and regulator is often undertaken in parallel with other project activities, such as investigation of electrical grid connection, detailed resource assessment and site specific development of an LCOE assessment. Public perception towards wave power developments is often more favorable than for onshore wind generation, due to the minimal visual impact of WEC arrays. On the other hand, the possible displacement of particular types of fishing at prospective development sites might result in concerns from the fishing community and fish processing businesses.

6.3 *Infrastructure constraints*

6.3.1 *Electrical grid connection*

Unless a WEC project development is planned for the desalination of water, or for direct use of the electricity at the project site, the connection to the electrical grid must be carefully considered. Depending on the project site location and its proximity to large population centers or other industrial users of energy, costs for electrical infrastructure can vary significantly. Often the most promising sites for WEC array installations are in exposed sparsely populated remote areas that feature the best wave resource. As an example, in the United Kingdom in 2012 more than 92% of wave power projects that had secured seabed leases from The Crown Estate were situated in the Scottish islands with a very limited local power use and insufficient electrical grid connection.[14] The electrical connection costs associated with both subsea cabling offshore and long distance HVDC transmission links onshore can be enormous, and this presents a challenge to project development in remote areas. The distance from a project site to shore, but also available grid capacity and cable length required to the next onshore transmission line, must therefore be evaluated at early project stages to maintain commercial feasibility.

6.3.2 *Geology*

Although the seabed geology is not a direct part of the project infrastructure, its composition and layout has a direct impact on the viability of the project. Depending on the technology specific requirements of a WEC installation, and the method used to connect devices to the seabed, geologic features might support or prevent a successful installation. Floating WEC devices are mainly secured at a location using single or multiple point mooring solutions. Where some devices require a catenary mooring system, with some damping provided by a slack mooring cable, other floating WECs are installed using a taut system with a permanent tension between float and ground weight. The type of mooring layout required, and whether embedment anchors, gravity-based clump weights or mechanical connections to the seabed can be used has considerable cost implications for procurement, installation and maintenance of the overall system. Therefore, an assessment of the type of seabed at a site is important to identify the depth of any sediment or gravel layers, or to evaluate the quality of bedrock as a possible solution for drilled anchor solutions. The existence of reefs, type and density of vegetation, or steepness and direction of slopes are all important considerations to enhance the understanding of wave attenuation and construction challenges. In addition to the geology directly at a project site, this must also be considered for possible cable routes to the shore, including suitable landfall sites for the electrical connection.

6.3.3 *Proximity to ports and harbors*

Depending on whether a WEC development is planned close to the coast in shallow water, directly on the shoreline or further offshore in deeper water, different approaches can be used for site development, construction and maintenance. Although for larger construction projects in deeper water it might be feasible to undertake the work very efficiently with dedicated offshore construction vessels at high charter costs, this is different closer to shore in a shallow water environment. In situations where larger vessels are constrained by depth, construction has to take place with smaller vessels, such as multi-category workboats. Due to crew sizes and operational limitations, these smaller craft often operate on a day-to-day basis from nearby ports or anchorages, and therefore the distance between a project site and a safe harbor must be considered. The same applies for the initial site development, which includes a large amount of survey and monitoring works, often done from small workboats during daylight hours only. On completion of the construction phase, it is further required to access the generation site for operations and maintenance purposes. Duration and costs for these operations also heavily depend on distance to the nearest port, and availability of suitable vessels. In cases where a WEC solution is to be installed on a harbor breakwater, or directly on the coastline, the requirement for workboat support might be limited, as most of the works can be undertaken from shore side. However, in those situations a careful assessment of the structural integrity of breakwaters is required to identify the load-bearing capability of the ground for craning operations, but also to ensure the impacting forces during WEC operations are within the limits of the carrying structure.

6.3.4 *Access to vessels (construction, operations, maintenance)*

As outlined in the previous section, different types of construction and support vessels may be required, depending on the site location and WEC technology. Although in the case of a large project it is likely feasible to invest in a dedicated vessel infrastructure for operations and maintenance, this is not a viable option for small one-WEC-only projects, or the different types of vessels required during the construction phase. In case of a dedicated vessel infrastructure from another sector, e.g. oil and gas, in relatively close proximity to a prospective WEC site, suitable construction vessels might be available for short periods at suitable times. This is the case in the north of Scotland, with Aberdeen being one of the main service bases for oil fields in the North Sea. In addition, due to the marine energy-related activity at the EMEC test site in Orkney, a considerable selection of smaller workboats of different specifications and sizes is available in the region, which makes this an attractive development area. Other areas don't have this benefit, with a requirement for construction vessels to spend various days in transit before any work can commence. In addition to the obvious cost implication of having to move vessels across longer distances at high charter costs, the absence of a local suitable vessel infrastructure also prevents the short-term scheduling of project activities to maximize use of suitable weather windows with favourable wind and wave conditions.

6.4   *Weather windowing*

Exposure to an energetic wave climate is an inherent feature of wave power development sites. Although this is highly desirable for energy production, challenges arise during the construction and maintenance periods. Dependent on the individual activities planned during the installation and operational phases, some tasks might only be carried out during times with limited wave action. Examples for this are the use of divers, or any lifting activities, where wave heights above a certain threshold pose a risk for safe operations due to the vessel's motion. To maximize efficient use of resources and therefore reduce the related costs, the ability to forecast the wind and wave situation at a project site is very important. This is referred to as 'weather windowing', where activities are planned for periods with forecasted benign conditions that allow for safe operations to take place as scheduled, without any unplanned downtime due to adverse weather conditions. In situations where vessels and resources are available near a project site, weather forecasting on a day-to-day basis might be sufficient. But if vessels have to be booked days in advance, due to availability or distance from the site, then it is desirable to identify suitable weather windows several days or weeks ahead of the operation. Therefore in areas with strong fluctuations of wave power between seasons, major operations are generally planned during the calmer months, i.e. in summer. A range of numerical wave models are available on large scale to support medium range weather windowing, with local or regional models and forecasts offering a suitable solution for short-term planning.

6.5 *Environmental impact*

Considerations of the environmental impact of a wave power installation are important to achieve and maintain societal acceptance and secure regulatory consent for development. This is analyzed during the environmental impact assessment (EIA), and generally categorized into biological, physical and societal impact as described in this section.

6.5.1 *Biological impact*

The potential interactions of a wave energy installation with the biological environment are of considerable concern and are widely investigated in an international approach. With 15 participating countries and initiated by the OES in 2010, the US DoE Annex IV program is likely the biggest initiative to identify and evaluate possible environmental impacts of marine energy installations.[15] Research related to the biological impact of wave farms considers both potentially positive and negative consequences of installations, although the qualification of an impact as good or bad is often subjective and dependent on specific evaluation criteria. A wave farm with floating devices might act as a platform for resting birds away from shore, thus allowing use of an extended foraging area. In addition, the lee side of energy converters might also provide some shelter on the sea surface supporting feeding and resting activities for birds. Depending on the type of protective coating of energy converters, varying degrees of biofouling will be present, and this can provide a food source and shelter for small fish. This presence of prey fish may attract larger predatory fish and marine mammals, and the possible increased abundance of fish and mammals within a wave farm is also supported by the absence of fishing activities close to the energy converters. As a result of an increased activity

of fish and birds near the surface close to the wave devices, the amount
of biomass in the form of dead fish, birds or algae also increases on the
seafloor, which may then attract lobsters or other crustaceans and bottom
feeders. At the same time, some displacement might take place of the initial
species that had occupied the habitat prior to the installation of the wave
farm. An increased abundance of marine mammals near a wave farm also
increases the collision and entanglement risks between animals and energy
converters and related infrastructure such as mooring cables. Depending
on the spacing between individual wave farms, these might act as 'stepping
stones' for certain species, with an increased spatial distribution as a result.
The biological impacts of marine energy installations are often separated
for marine mammals (including basking sharks), fish, birds and the benthic
environment, with individual assessments required for each species. Impact
assessments consider both the construction and operational phases, with
consideration given to issues such as temporary noise pollution (e.g. during piling operations during the construction phase), which might lead to
a temporary displacement of species. The impact of leaking hydraulic systems or other forms of pollution are also part of the assessment. Figure 8
shows some Arctic Terns roosting on a Pelamis P2 energy converter at the

Fig. 8.   Roosting Arctic Terns on Pelamis WEC at EMEC test site 2013.[16]

EMEC test site in Orkney, Scotland, in 2013. Also visible in the picture is the breakwater effect of the wave attenuator, with calmer water on the down-wave site.

6.5.2 *Physical impact*

Less attention is given to the physical impact of wave farms when compared with possible interactions with the biological environment. One area of concern is the reduction of wave power for additional energy installations further down wave. Due to the absence of large wave energy converter installations to date, this is presently assessed using numerical computer models, without the possibility of validation of such models against measured data. Some concern is expressed by the surfing community, as the large-scale extraction of wave power might result in a reduced quality and quantity of breaking waves in the surf zone. The potential of wave energy extraction to reduce the amount of wave power absorbed by the shoreline, and thus to alleviate coastal erosion or flooding, is generally seen as a positive impact from a coastal protection point of view.

6.5.3 *Societal impact*

The possible impact of wave energy installations on society is mainly seen in the interaction with offshore users, as project sites might be closed to marine traffic and fisheries. In addition to the displacement of fishing activities directly at the sites, further impact might also be seen in the reduced availability of vessel berthing space in nearby harbors. This is due to the requirement for service and crew transfer vessels related to the wave farms, which compete for often limited space in harbors in exposed remote areas. Concerns of the surfing community on the impact of wave farms on the quality of surf have already been mentioned in the previous section, and this might have an impact on the extent of water sport tourism close to wave energy sites. Another consideration is the visual impact of wave farms, which is typically small and always much less when compared with wind farms. But visual impact also refers to any required onshore infrastructure, including electrical transmission cables, and depending on the installation area, this might be perceived as a negative intrusion. But overall, the potential positive impact of wave energy developments on the local economy is seen to largely outweigh any possible negative implications, provided local concerns of other users of the marine environment are taken into account at an early planning stage.

7   Outlook

Following a major setback to the wave energy sector around 2014, when the then world leading WEC technology companies Wavegen, Pelamis and Aquamarine ceased trading, more recent developments have resulted in a positive anticipation of future activities. Largely driven by public sector initiatives such as the DoE Wave Energy Prize or Wave Energy Scotland, momentum has been regained, and currently a considerable number of technology developers are at the stage of prototype testing of complete WEC devices or at the component level. Current test programs extend from small-scale device studies on laboratory tank scales, up to full-scale energy converters in real sea environments. Efforts to bring the wave power sector to commercial market readiness are well underway around the globe, often in intergovernmental programs or in a collaborative approach between transnational acting individual developers. The European Union Blue Growth program continues to invest in research and prototype testing of wave energy converters as dedicated standalone solutions, or in combination with other existing offshore structures. Very recently in 2019, the US DoE announced a competition to develop wave powered water desalination systems,[17] an aim that is also being pursued in China and many other countries and by different developers.

The experiences gained in WEC prototype and component testing, site development and detailed resource characterization over the recent decade have created a strong foundation for further development of the sector, and currently it seems indicated that the willingness of private sector investors to provide funds for project developments is increasing.

Although it is not possible with absolute certainty to predict how the wave energy sector will develop in the future, the extensive global search for low carbon energy solutions and water security, together with the recent promising advances made in the development of WEC technology, provides a strong foundation and positive outlook for the wave energy sector.

References

1. G. Mørk, *et al.* (2010). Assessing the global wave energy potential. In *Proceedings of OMAE2010 (ASME)*, *29th International Conference on Ocean, Offshore Mechanics and Arctic Engineering*. Shanghai, China.
2. World Energy Council (2016). World energy resources 2016. London. https://www.worldenergy.org/wp-content/uploads/2016/10/World-Energy-Resources-Full-report-2016.10.03.pdf.

3. Ocean Energy Systems (2017). Annual report — An overview of activities in 2017. Edited by e Melo, A. B. and Jeffrey, H. Lisbon: *The Executive Committee of Ocean Energy Systems*, p. 4.

4. Ocean Energy Systems (2016). 2016 Annual Report. Edited by e Melo, A. B. and Villate, J. L. Lisbon: *The Executive Committee of Wave Energy Systems*, p. 6.

5. BSI (2015). Marine energy — Wave, tidal and other water current converters, Part 101: Wave energy resource assessment and characterization. London: The British Standards Institution. PD IEC/TS 62600-101:2015.

6. Department of Industry (1976). The development of wave power. *A Techno-Economic Study. Glasgow: National Engineering Laboratory*, EAU M25.

7. M. Seed and D. Langston (2010). Wave energy — Towards commercialisation. *In: Proc. of the 3rd Intern. Conf. on Ocean Energy (ICOE)*, Bilbao, October 2010. https://www.icoe-conference.com/publication/wave_energy_t owards_commercialisation/.

8. https://www.ocean-energy-systems.org/index.php.

9. https://www.oceanenergy-europe.eu/.

10. H. Jeffrey (2018). IEA technology collaboration programme Ocean Energy Systems (TCP OES). *Wave Energy Scotland annual conference 2018*, Edinburgh.

11. Carnegie Clean Energy (2018). CETO 6 — Albany (WA). https://www.ca rnegiece.com/project/ceto-6-albany/.

12. Catapult ORE (2018). Tidal stream and wave energy cost reduction and industrial benefit: Summary analysis. Glasgow, Blyth and Levenmouth: Catapult ORE.

13. European Commission (2018). Market study on ocean energy: Final Report, ISBN: 978 92 79 87879 4. Brussels: European Commission.

14. Aquamarine Power Ltd (2013). The locational drivers of wave energy projects Accessed 09/12/2013, URL closed.

15. OES Annex IV (2010) About Annex IV https://tethys.pnnl.gov/about-ann ex-iv.

16. Hebridean Marine Energy Futures (2014). Workstream 4. Onboard camera PELAMIS device, University of the Highlands and Islands.

17. US Department of Energy (2019). DoE announces prize competition wave energy water desalination. https://www.energy.gov/articles/doe-announces -prize-competition-wave-energy-water-desalination.

Chapter 8

Tidal Power

Sue Molloy

Glas Ocean Electric, Halifax, Nova Scotia, Canada
Sue@glasocean.com

Abstract

Power generated from the tides is a predictable renewable energy source. Tidal power using the flow of tides rather than the drop in water from a height means that structures like dams are unnecessary. Coastal regions are familiar with tides and measuring the rise and fall of the waterline has been a part of living on the coast for centuries. When ensuring the environmental impact is managed thorough adaptive management, the tides offer the opportunity to build a sustainable base-load power for renewables. Projects around the world have demonstrated the potential of tidal power and designs are beginning to converge on a small number of concepts.

1 Introduction

Power generated from the tides is a predictable renewable energy source. Using the tides offers the opportunity to build a sustainable base-load power for renewables and when used in combination with storage, wind and/or solar 100% renewable sustainable grids become a possibility outside of hydropower.

All coasts have some kind of tidal flow and some coasts have enough flow to be able to consider extraction of that energy to power entire communities.

2 What are Tides?

Coastal regions are familiar with tides and measuring the rise and fall of the waterline has been a part of living on the coast for centuries. This alternating rising and falling of water bodies has also been measured inland on lakes although with much smaller changes.

Tides are the localized change in water depth of oceans and other water bodies caused by the gravitational pull of the Moon and the Sun. The gravitational force of the Moon is substantially higher than the force of the Sun because the tide generating force is proportional to the mass divided by the distance cubed.[1]

$$Tide\ Generating\ Force \propto \frac{Mass}{Distance^3} \tag{1}$$

Essentially, while the Sun has 27 million times more mass than the Moon, it is 390 times further away, which means the Moon impacts the tide more than the Sun.

> **Does the land move as well as the sea?**
> *Yes! In very small but measurable amounts. Encyclopedia Brittanica calls these earth or land tides.[2]*

The Moon is the defining force in this relationship and tides follow the lunar day. Because the Moon is continuously orbiting the earth, the total tide force changes in magnitude through the lunar month. The Sun provides an additive gravitational force.

This means that the power available for capture in the tidal flow changes with the position of the Moon throughout the lunar month, see Fig. 1. The positions of the Moon are generally referred to as full, new and quarter. When the Moon is moving to full, it is said to be waxing, when it is moving out of full Moon, it is waning. A great scrabble word is the one used to describe when the Moon, earth and Sun are in a straight line: Syzygy.

The neap tide is the smallest tide because the Moon is moving from the 3rd quarter to the new Moon and the additive effect of the Sun and Moon together is at its minimum. This means that the tidal flow speed is at its lowest for the month at neap tide. Tidal power installations and operations are usually planned for the neap tide window so the vessels required are working in slower flow.

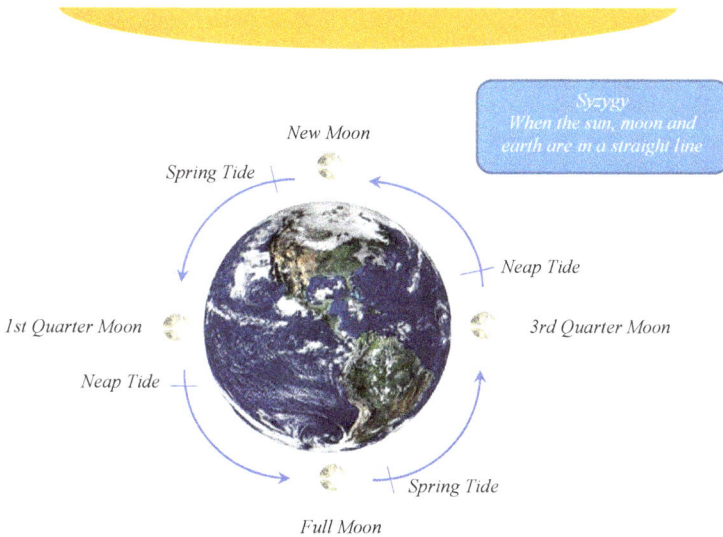

Fig. 1.   Moon positions around the earth.

A lunar day, defined here as moonrise to moonrise as viewed from earth, lasts 24 hours and 50 minutes because the Moon revolves around the earth in the same direction that the earth rotates around its axis. The tides then shift in time daily, which means although the tides are predictable there is significant effort involved in combining the positions of the Moon and Sun with historical knowledge, locations, depth, bathymetry and more to model tidal currents and heights and accurately predict the time and flow of tidal height changes.

3   Tide Tables

The tides for any particular day at a particular location can be found at a resolution of ∼6 hours in government tide tables. A sample of the tide table for Halifax, Nova Scotia, for January–March 2019 is found in Fig. 2.

The information in the table includes the location, the time zone, the date, the time, the expected height in meters and feet. The change in the height with time is shown in Fig. 3. This is the preferred data for a tidal site from tidal power developers, but it is not always available publicly. Developers and researchers collect this type of data for sites using equipment such as flow meters, lidar radar and acoustic current profilers.

HALIFAX AST Z+4   **2019**   TIDE TABLES

January-janvier

Day	Time	Metres	Feet	Day	jour heure	mètres	pieds
1 TU MA	0400 1059 1640 2314	1.7 0.3 1.5 0.4	5.6 1.0 4.9 1.3	**16** WE ME	0256 0953 1538 2153	1.7 0.5 1.5 0.6	5.6 1.6 4.9 2.0
2 WE ME	0455 1152 1738	1.7 0.3 1.6	5.6 1.0 5.2	**17** TH JE	0354 1049 1644 2251	1.7 0.4 1.5 0.5	5.6 1.3 4.9 1.6
3 TH JE	0008 0545 1241 1829	0.5 1.7 0.2 1.6	1.6 5.6 0.7 5.2	**18** FR VE	0451 1145 1743 2350	1.8 0.2 1.6 0.5	5.9 0.7 5.2 1.6
4 FR VE	0059 0631 1326 1916	0.5 1.7 0.2 1.7	1.6 5.6 0.7 5.6	**19** SA	0547 1240 1838	1.9 0.1 1.7	6.2 0.3 5.6
5 SA SA	0144 0715 1408 1959	0.5 1.8 0.2 1.7	1.6 5.9 0.7 5.6	**20** SU DI	0048 0642 1334 1931	0.4 2.0 0.0 1.8	1.3 6.6 0.0 5.9
6 SU DI	0224 0758 1445 2042	0.6 1.8 0.3 1.7	2.0 5.9 1.0 5.6	**21** MO LU	0145 0737 1427 2023	0.3 2.0 -0.1 1.9	1.0 6.6 -0.3 6.2

February-février

Day	Time	Metres	Feet	Day	jour heure	mètres	pieds
1 FR VE	0525 1218 1812	1.6 0.3 1.6	5.2 1.0 5.2	**16** SA SA	0421 1121 1719 2331	1.8 0.3 1.6 0.5	5.9 1.0 5.2 1.6
2 SA SA	0038 0614 1304 1859	0.5 1.7 0.3 1.6	1.6 5.6 1.0 5.2	**17** SU DI	0527 1219 1818	1.9 0.1 1.7	6.2 0.3 5.6
3 SU DI	0123 0659 1345 1941	0.6 1.7 0.3 1.7	2.0 5.6 1.0 5.6	**18** MO LU	0033 0626 1314 1913	0.4 2.0 0.0 1.8	1.3 6.6 0.0 5.9
4 MO LU	0200 0740 1422 2020	0.6 1.8 0.3 1.7	2.0 5.9 1.0 5.6	**19** TU MA	0132 0723 1317 2005	0.3 2.0 -0.1 1.9	1.0 6.6 -0.3 6.2
5 TU MA	0232 0820 1454 2056	0.6 1.8 0.3 1.7	2.0 5.9 1.0 5.6	**20** WE ME	0230 0816 1458 2055	0.2 2.0 -0.1 2.0	0.7 6.6 -0.3 6.6
6 WE ME	0302 0858 1524 2131	0.6 1.8 0.3 1.7	2.0 5.9 1.0 5.6	**21** TH JE	0327 0908 1549 2144	0.2 2.0 -0.1 2.0	0.7 6.6 -0.3 6.6

March-mars

Day	Time	Metres	Feet	Day	jour heure	mètres	pieds
1 FR VE	0359 1059 1656 2323	1.5 0.4 1.5 0.6	4.9 1.3 4.9 2.0	**16** SA SA	0242 0958 1546 2216	1.7 0.4 1.5 0.6	5.6 1.3 4.9 2.0
2 SA SA	0504 1150 1751	1.6 0.4 1.6	5.2 1.3 5.2	**17** SU DI	0359 1059 1700 2320	1.8 0.3 1.6 0.5	5.9 1.0 5.2 1.6
3 SU DI	0013 0555 1236 1836	0.6 1.6 0.4 1.6	2.0 5.2 1.3 5.2	**18** MO LU	0511 1157 1800	1.9 0.2 1.8	6.2 0.7 5.9
4 MO LU	0056 0639 1317 1916	0.6 1.7 0.3 1.7	2.0 5.6 1.0 5.6	**19** TU MA	0021 0613 1253 1853	0.3 1.9 0.1 1.9	1.0 6.2 0.3 6.2
5 TU MA	0132 0719 1352 1952	0.5 1.7 0.3 1.7	1.6 5.6 1.0 5.6	**20** WE ME	0120 0708 1345 1943	0.2 2.0 0.0 2.0	0.7 6.6 0.0 6.6
6 WE ME	0203 0757 1423 2026	0.5 1.8 0.3 1.7	1.6 5.9 1.0 5.6	**21** TH JE	0216 0800 1435 2030	0.1 2.0 0.0 2.0	0.3 6.6 0.0 6.6

Fig. 2.   Tide table sample for Halifax, Nova Scotia, Canada.[3]

When considering extracting energy from tidal sites, there are three aspects of the tide that are measured. The height, the current and the frequency of change. Some sites are diurnal with one high tide and one low tide approximately every 24 hours, others are semi-diurnal with 2 high tides and 2 low tides approximately every 24 hours.[5]

In general, a tidal site with a significant change in height can be expected to have a high-speed current because in order to change height within a 6-hour window, the water must move fast. A sample current table for the Great Bras d'Or region is shown in Fig. 4. The table shows the location, time zone, date, time, speed in knots and direction.

The change in current with tide is illustrated in Fig. 5 from the University of Hawaii. When tide is "coming in" or the water is rising, it is called a flood tide, when the tide is "going out" or the water depth is lowering, it is called an ebb tide. In tide tables, the direction is identified by − or + where "−" is an ebbing tide and "+" is a flooding tide.

Tidal depths in tables and charts are measured from the chart datum, which is the depth or level of water that the chart is measured from, often the lowest astronomical tide, not the sea floor.

4   Tide Location

Tides change with location. The shape and depth of bays can impact the timing and speed of depth change. This offers a potentially exciting

Distribution of Tidal Phases

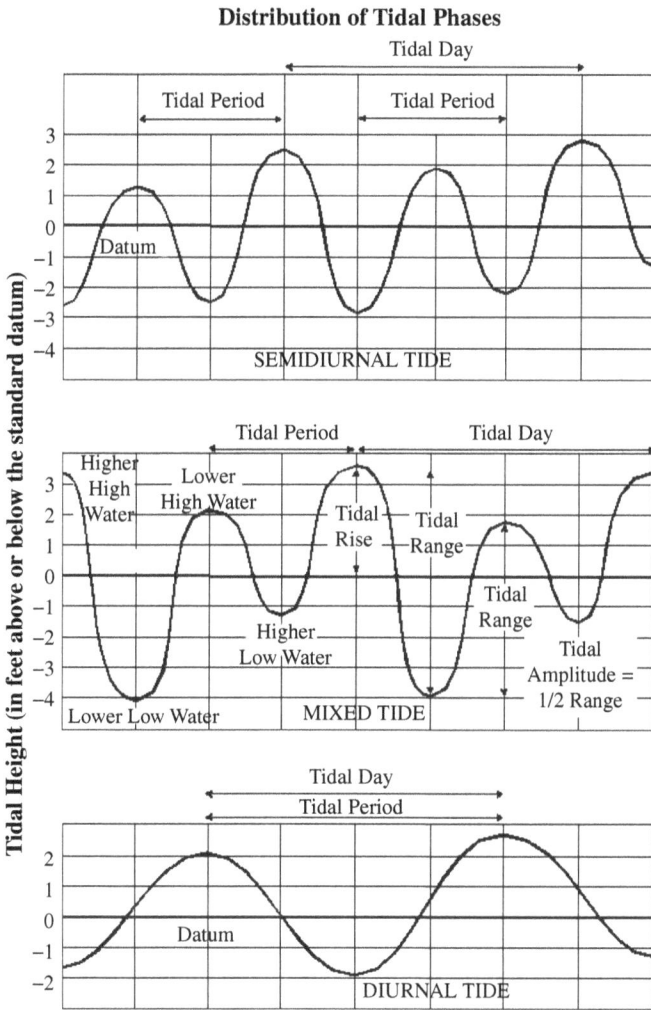

Fig. 3.   Tidal heights.[4]

opportunity for extracting power from the tides. In a country, state or province, it is possible that the timing of the tide can shift over hours along the coast.

This means that:

If tidal devices are staggered and optimized along a coastline, there could be continuous power from tides

GREAT BRAS D'OR AST Z+4   2019   CURRENT TABLES

January-janvier

Turns (Day Time)	Maximum (Time Knots)	renverse (jour heure)	maximum (heure noeuds)
1 0234	0515 +2.3	16 0151	0421 +1.5
0813	1112 -2.5	0709	0953 -1.6
TU 1424	1736 +2.8	WE 1247	1556 +2.9
MA 2037	2352 -3.3	ME 1940	2253 -2.9
2 0342	0630 +2.2	17 0308	0536 +1.4
0918	1206 -2.4	0820	1056 -1.6
WE 1511	1824 +3.0	TH 1342	1656 +3.2
ME 2124		JE 2038	2351 -3.3
3 0443	0046 -3.6	18 0413	0653 +1.6
TH 1014	1254 -2.4	FR 0926	1158 -1.9
JE 1554	1905 +3.1	VE 1443	1757 +3.5
2209		2133	
4 0537	0131 -3.8	19 0506	0042 -3.6
FR 1102	0826 +2.4	SA 1026	0751 +2.0
VE 1634	1336 -2.4	SA 1545	1256 -2.2
2252	1943 +3.3	2224	1854 +3.8
5 0623	0212 -3.8	20 0551	0129 -3.9
SA 1143	0905 +2.4	SU 1123	0837 +2.6
SA 1711	1415 -2.5	DI 1644	1350 -2.6
2334	2020 +3.4	2313	1947 +4.0
6 0701	0252 -3.7	21 0633	0216 -4.2
SU 1219	0937 +2.4	MO 1216	0919 +3.1
DI 1749	1454 -2.7	LU 1739	1440 -2.8
	2057 +3.5		2037 +4.1

February-février

Turns (Day Time)	Maximum (Time Knots)	renverse (jour heure)	maximum (heure noeuds)
1 0440	0031 -3.4	16 0359	0637 +1.4
0959	0727 +1.9	SA 0906	1137 -1.7
FR 1513	1225 -1.9	SA 1417	1740 +3.3
VE 2147	1841 +3.0	2107	
2 0531	0120 -3.5	17 0449	0024 -3.6
SA 1045	0813 +2.0	SU 1013	0735 +2.1
SA 1559	1310 -2.0	DI 1530	1242 -2.2
2234	1924 +3.2	2203	1845 +3.6
3 0611	0201 -3.5	18 0532	0114 -3.9
SU 1124	0848 +2.1	MO 1110	0819 +2.7
DI 1644	1351 -2.3	LU 1636	1336 -2.6
2318	2004 +3.4	2256	1940 +3.8
4 0641	0237 -3.4	19 0615	0203 -4.2
MO 1159	0915 +2.3	TU 1200	0858 +3.2
LU 1727	1431 -2.6	MA 1733	1424 -3.0
2357	2041 +3.5	2349	2029 +4.1
5 0705	0310 -3.4	20 0657	0253 -4.3
TU 1235	0940 +2.5	WE 1245	0935 +3.4
MA 1810	1510 -2.8	ME 1825	1508 -3.3
	2116 +3.5	2337	2117 +4.2
6 0032	0341 -3.4	21 0041	0344 -4.4
0727	1005 +2.8	0738	1012 +3.4
WE 1313	1549 -2.9	TH 1326	1552 -3.5
ME 1852	2151 +3.4	JE 1916	2207 +4.2

March-mars

Turns (Day Time)	Maximum (Time Knots)	renverse (jour heure)	maximum (heure noeuds)
1 0326	0603 +1.3	16 0233	0453 +1.2
0842	1101 -1.4	0732	1007 -1.4
FR 1339	1722 +2.5	SA 1247	1614 +2.8
VE 2028		SA 1940	2305 -3.3
2 0427	0009 -3.1	17 0333	0618 +1.7
0942	0711 +1.6	0856	1128 -1.8
SA 1439	1157 -1.5	SU 1415	1736 +3.0
SA 2123	1816 +2.8	DI 2045	
3 0513	0100 -3.2	18 0423	0009 -3.6
SU 1028	0755 +1.8	MO 0959	0714 +2.3
DI 1534	1245 -1.8	LU 1532	1230 -2.4
2213	1902 +3.0	2146	1839 +3.3
4 0547	0141 -3.2	19 0509	0102 -3.9
MO 1105	0825 +2.0	TU 1051	0757 +2.9
LU 1626	1328 -2.2	MA 1635	1321 -2.9
2258	1944 +3.3	2243	1932 +3.7
5 0611	0215 -3.2	20 0552	0151 -4.1
TU 1137	0848 +2.3	WE 1135	0834 +3.2
MA 1715	1409 -2.6	ME 1730	1405 -3.3
2337	2023 +3.4	2337	2020 +4.0
6 0631	0246 -3.2	21 0633	0241 -4.2
1209	0907 +2.6	0906 +3.3	
WE 1209	1447 -3.0	TH 1215	1446 -3.7
ME 1801	2059 +3.4	JE 1820	2108 +4.1

Fig. 4.   Current tables for the Great Bras d'Or Region.[3]

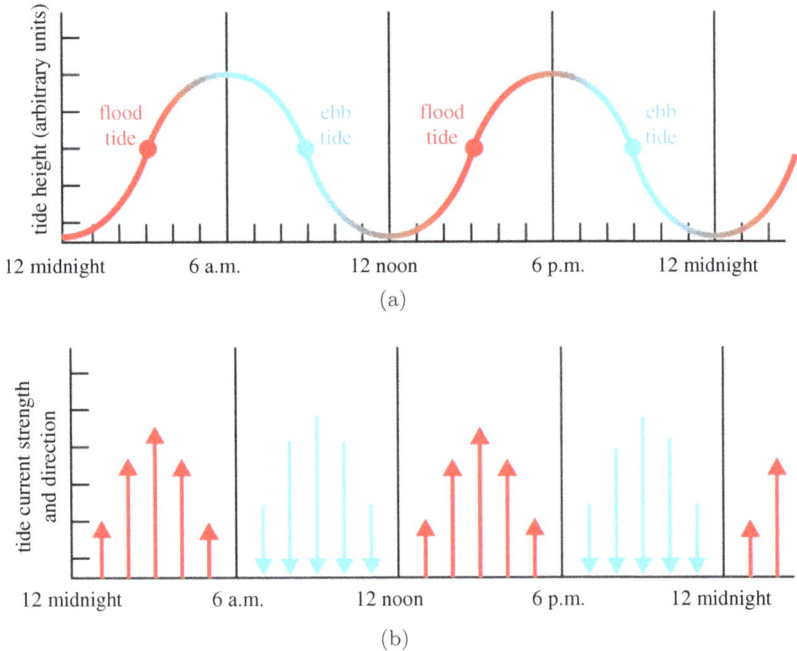

tide height (arbitrary units)

flood tide   ebb tide   flood tide   ebb tide

12 midnight   6 a.m.   12 noon   6 p.m.   12 midnight

(a)

tide current strength and direction

12 midnight   6 a.m.   12 noon   6 p.m.   12 midnight

(b)

Fig. 5.   Tide height and tidal current change direction and strength throughout the day in one location. Longer arrows indicate stronger tidal currents, Graphic from Byron Inouye, University of Hawaii.[6]

Fig. 6. Fishing boat on the ocean floor at low tide in the Bay of Fundy, photo credit: Sue Molloy.

Some locations have a combination of bathymetric and geologic coastline features that create an amplification of tides. This causes "resonance" and leads to very high tides. An example of this is in the Bay of Fundy in Canada. See Fig. 6, where tide heights at maximum in some locations are over 16 m.

Another unique feature of sites with exceptionally high tides is that because the water has to move so fast twice per day to fill or empty the space, there are often shear forces in the water column due to water still flooding when the ebbing begins. These shear forces can have impacts on the installation, operation and maintenance of tidal power systems and can cause the device to be subjected to very high loads.

5 Predictability

As seen from the tidal tables, the predictability of tides is what makes tidal power so attractive. By having predictable renewable power paired with renewables such as wind or solar and adding in storage, there is potential for a 100% renewable and sustainable grid. Optimizing the power extraction opportunity by staggering the location of tidal devices means that the storage component would not need to be 100% of the maximum required power. A combination of storage and tidal power to "ride the micropeaks" or stabilize the power variations caused by solar and wind variations throughout the day could be very appealing to power generation companies. Currently, this variation in power can be managed by 100% energy storage or diesel generator backup. Diesel generator power is often US$0.40/kWh or more.

When a full accounting of the environmental, health and safety and climate change impacts of the diesel systems are considered, the tidal/storage combination becomes very attractive.

6 In-Stream Versus Tidal Barrage

Many people understand tidal power as power which is extracted from the change in height of the water. This primarily describes tidal barrages. This system dams the water in a reservoir or head pond as the tide floods and then later as the tide ebbs the water is released slowly through hydro-turbines to generate power.[7]

These barrages have a significant height difference so that the power is generated by the energy of the water in a way similar to a hydroelectric dam. There are very few tidal barrage systems in operation; Annapolis, Nova Scotia, Canada, Fig. 7, and La Rance, France, are the most well-known installations.

In-stream tidal power is best described as being like wind turbines under the water, where the flow is of water rather than air. In-stream turbines use the energy in the flow of the water to turn rather than the change in height. A site with a large tidal height is a premium in-stream

Fig. 7.   Annapolis tidal power station, Nova Scotia, Canada, photo NovaScotia.com.[8]

site because, as mentioned before, filling in a large space quickly means the water has to flow fast. This can be seen as being like a site with high winds.

7 Resources

The sites with the highest tides, listed in Table 1, have been considered and some used for tidal power installations. The Bay of Fundy in Canada has a barrage installation and currently there are a number of in-stream tidal projects being developed.

In-stream tidal power is considered a larger and more viable opportunity than barrage because the number of turbines can be scaled to fit a site and a location does not need to be dammed which is a significant environmental challenge for these locations that feed inter-tidal zones. The sites identified as suitable for tidal power around the world are included on the map in Fig. 8. Estimates by Charlier and Justus (1993) of the amount of tidal power that can be extracted worldwide are in the range of 3TW.[11]

These estimates indicate the potential resources; however, the accessibility of the locations, the availability of marine assets and construction resources, the distance to the grid and more have a direct impact on the ability to extract the resource.

8 Micro-Siting

Using a map such as that in Fig. 8 can help developers determine where to consider siting a tidal power project, but the details on the ground need to be well-understood before making any final decisions. The local environment consists of at minimum:

- bathymetry;
- geology;
- flow/current speed changes through the day;
- flow/current direction;

Table 1. Location with highest tides globally.[9]

Location	Country	Approx. Max Range (m)
Bay of Fundy	Canada	16.2
Ungava Bay	Canada	16
Port of Granville	France	14.7
Severn Estuary	UK	14.5
La Rance	France	13.5
Puerto Rio Gallegos	Argentina	13.5

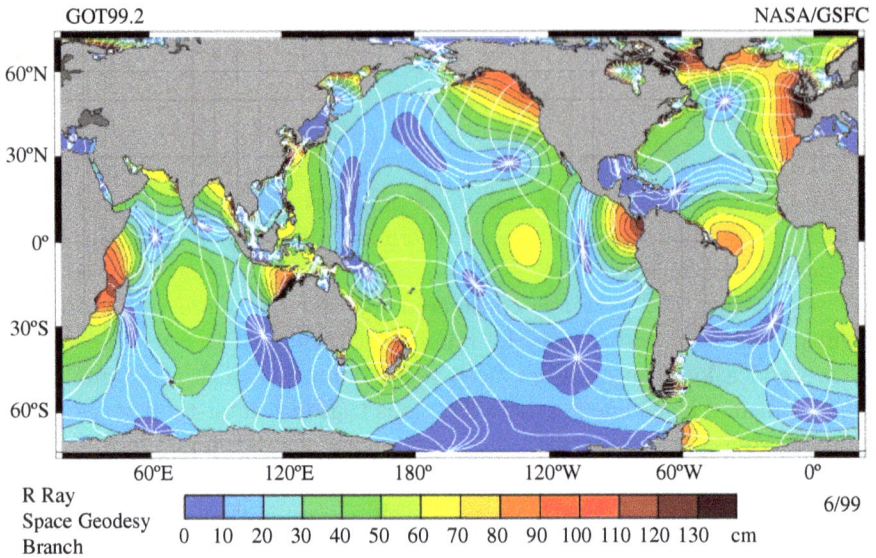

Fig. 8.   High potential sites for tidal power, graphics from NASA.[10]

- amount and quality of turbulence in the flow;
- changes in flow/current speed with depth;
- sedimentation in the water;
- animals sharing the space;
- distance to shore;
- distance to grid.

These details help the developer to decide on the installation design. Obstacles identified by bathymetry studies such as underwater ridges can change the flow, and speed up or slow down in narrow channels can put uneven flow over the turbine. The geology helps identify the best mooring plan: for example, if the turbine is mounted directly to the sea floor, the ability to drill into the floor needs to be determined. The tide sometimes flows at different angles when it is flooding compared to when it is ebbing.

The Annual Energy Production (AEP) needs to be properly predicted to determine if the cost of developing the site is justified. AEP is directly related to the size of the turbine installed, the amount of time the turbine will be operating at its rated capacity and in particular to the turbulence in the flow and the quality of the flow. The type and size of the loads expected on the system will determine if a floating or floor-mounted system would be required. Sedimentation can wear the turbine components and become unsettled by the turbine, impacting the marine environment. Animal habitats that can be impacted need to be understood to ensure appropriate mitigation. The ability to transmit or distribute the power that is generated is very important for the overall project; power transmission lines can be very expensive, >$1M/km.

All of these aspects are being studied around the world in companies, academia and government research institutes. Any information that can help to further understand tidal sites will increase the potential for success of projects. Our general understanding of these sites is very limited; it has been said we know more about space than the ocean, and while that is clearly not possible (we do not yet know the extent of the universe!), it does make the point that we are only beginning to learn about the ocean.[12]

Areas of focused study include:

- Turbulence — a highly turbulent site is actually desirable if the turbulence is in fully formed flow, because this turbulence is the energy that gets converted to electricity.
- Drilling — ways that drill ships can be used and maintain a station in high flow.
- Animals — the variety and habitats of animals in potential sites are studied to determine how they could be impacted by turbines.

9 Power

When determining how much power a device can extract from a given tidal flow, the power equation (2) is the starting point.

$$Power = \frac{1}{2}\dot{m}V^3 \tag{2}$$

where the power is that which can be captured from a fluid with a velocity V and a mass flow rate $\dot{m}$ that is flowing over an area A.

And the mass flow rate, $\dot{m}$, is defined as

$$\dot{m} = \rho A V \tag{3}$$

where ρ is the density of the fluid.

To measure the maximum power that is available to be extracted from a fluid by an area A, the $\dot{m}$ is substituted into the power equation to give

$$Power = \frac{1}{2}\rho A V^3 \tag{4}$$

This is the classic power equation used in both the wind industry and the tidal turbine world.

$Power_{Maximum}$ is the maximum power that is in the flow. This does not account for losses due to the shape of the turbine, friction, etc. As is intuitively understood, some turbine designs produce more power than others. The efficiency of the turbine is the ratio of the power actually extracted to the maximum available power.

$$C_p = \frac{Power_{Extracted}}{Power_{Maximum}} \tag{5}$$

When representing the power of a particular turbine, the $Power_{Maximum}$ equation is multiplied by the efficiency of that turbine to give the maximum power the turbine could deliver.

$$Power_{Extracted} = C_p Power_{Maximum} \tag{6}$$

where C_P is the power coefficient of the individual turbine representing how efficiently that turbine extracts power. This is a non-dimensional value that allows comparison of different turbines. Turbine designers determine their efficiencies through modeling and experimentation. Industry standard organizations such as the International Electrotechnical Commission (IEC) work to provide standards that ensure the efficiencies and capacities of turbines are being estimated and shown in consistent and reliable ways.

A traditional way of representing the performance of a wind or tidal turbine is the "C_P versus TSR plot", as shown in Fig. 9. This is the efficiency of the tidal turbine versus the tip speed ratio, TSR. The tip speed ratio is the ratio of the rotational speed at the tip of the turbine blade to the fluid speed. That is, the speed the turbine rotating measured at the outer diameter versus the wind or water flow speed.

Hodge[9] has a plot in his textbook on energy systems that shows the differing efficiencies of various wind turbines (see Fig. 10). This helps developers determine which turbine would be best for the flow regime in the site they have chosen. Additionally, the cost of some turbines is significantly higher than others based on their design, materials, components and installation requirements. This plot can help differentiate the turbine options. While there is not yet a comprehensive similar plot for tidal turbines, the wind turbine plot provides useful information along with data from individual turbine developers.

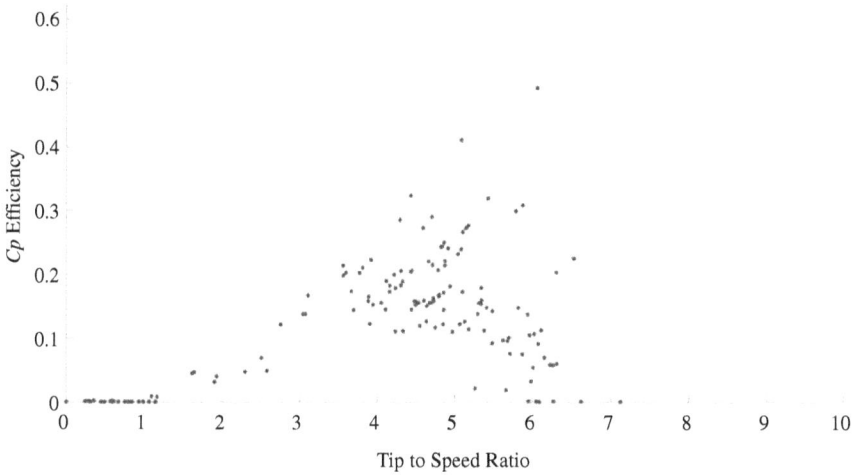

Fig. 9.   Sample C_P versus TSR plot from a tidal turbine test.

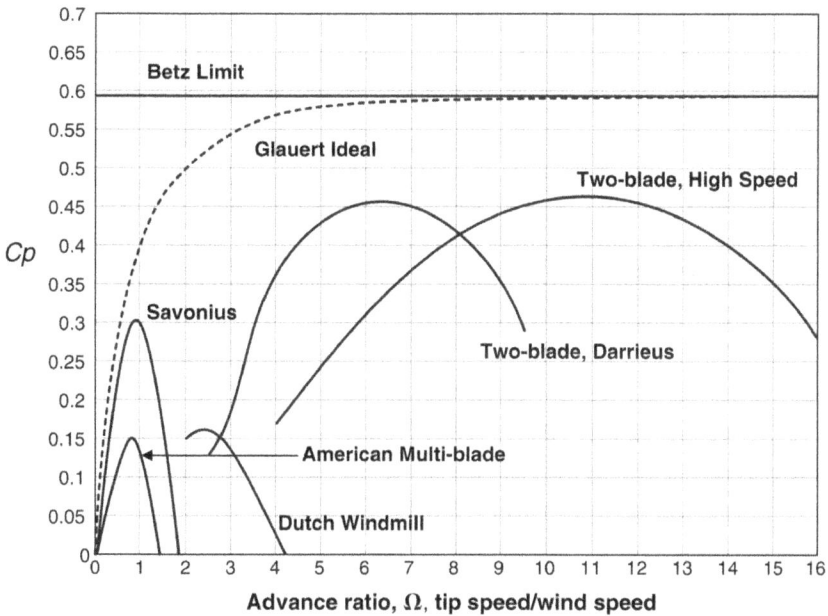

Fig. 10. C_P versus advance ratio for Wind Turbines, Hodge (2010).

The Betz Limit wind power theory[9] states that the maximum efficiency that can be achieved by a wind turbine is ~59%. This limit is considered to also apply to tidal turbines.

10   Comparing Wind and In-Stream Tidal Power

A valuable way to describe the amount of power in tidal flow is to show the equivalent power in wind. For some, a tidal flow of 5 km/hr does not intuitively seem like high speed.

Using the power equation from Equation (6),

$$Power_{ext} = \frac{1}{2} C_p \rho A V^3_{wind}$$

Assume we have turbines that are the same diameter/area and are required to produce the same power in water and wind. We will input values to determine what the water speed would need to be to achieve the same power in the tidal turbine as the wind turbine.

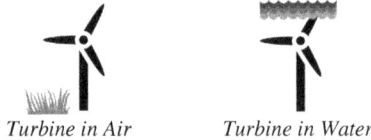

Turbine in Air   *Turbine in Water*

Two turbines, having the same diameter and same surface area, both produce the same power with the same efficiency.

To keep it simple, we will assume the wind speed is 100 km/hr

Density of Air $\rho_A = 1.22$ kg/m^3

Density of Seawater $\rho_W = 1027$ kg/m^3

$$Power_{Wind} = C_P \frac{1}{2} \rho A V^3 = Power_{Tidal}$$

$$C_P \frac{1}{2} \rho_{Air} A V^3_{Air} = C_P \frac{1}{2} \rho_{Water} A V^3_{Water}$$

$$\rho_{Air} V^3_{Air} = \rho_{Water} V^3_{Water}$$

$$V_{Water} = \sqrt[3]{\frac{\rho_{Air} V^3_{Air}}{\rho_{Water}}}$$

Hence, for a wind velocity of 100 km/hr, the water speed to produce the same power with the same turbine is 10.6 km/hr. As a general rule, if a turbine is in a water flow, that flow is equivalent to approximately 10 times that speed in air.

Flow speeds in the Bay of Fundy are as high as 5 m/s, which is 18 km/hr. This means that the power that can be produced in the Bay of Fundy using a tidal turbine is equivalent to that produced by the same turbine in 180 km/hr hurricane force winds.

11   Design of Tidal Turbines

There are a number of types of tidal turbine. *Horizontal axis* and *vertical axis* are the most common styles developed; *cross-flow* and *Archimedes screw* styles have also been tested.

In horizontal or axial-turbines, the rotational axis of the rotor is parallel to the inflow and is fitted with lift or drag type blades. In vertical turbines, the rotational axis of the rotor is vertical to the water surface and orthogonal to the inflow, again with lift or drag blades.

Figure 11 (i) shows the kinds of horizontal axis turbines that have been designed. The rigid mooring of (i)(b) is a typical ocean floor mounted style that has been tried in a number of sites around the world, (i)(d) the floating style, is the other most commonly tested design. These turbines also have different styles of blades including 2 blade, 3 blade, 4 blade and open center. The vertical axis turbines (b) are more varied in style and the devices are primarily floated from a platform or jetty/pier. Ocean floor mounted designs have been proposed but not yet tested.

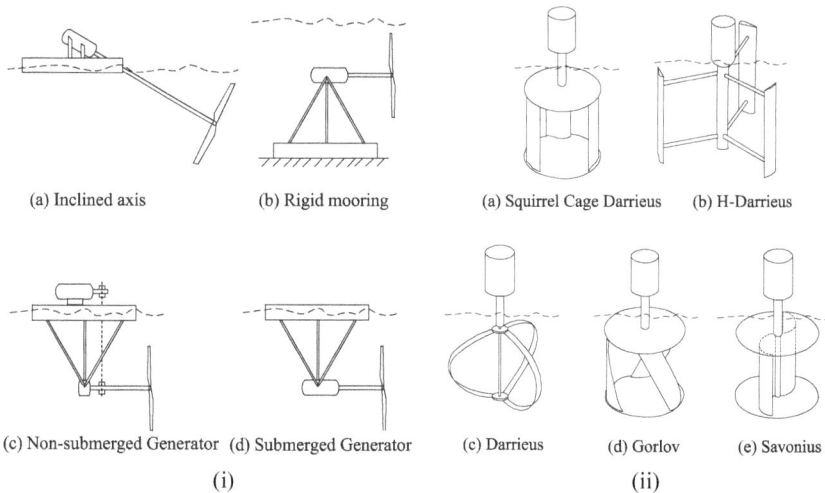

(a) Inclined axis   (b) Rigid mooring   (a) Squirrel Cage Darrieus   (b) H-Darrieus

(c) Non-submerged Generator (d) Submerged Generator   (c) Darrieus   (d) Gorlov   (e) Savonius

(i)   (ii)

Fig. 11.   Types of Tidal Turbines (a) Horizontal Axis, (b) Vertical Axis, Kahn *et al.* (2009).

Fig. 12.   SCHOTTEL Tidal Turbine, graphic reprinted with permission.

An example of a commercial horizontal axis turbine is shown in Fig. 12. This turbine is one of the Schottel turbines from Germany and has been in operation in Europe and Canada.

Horizontal axis turbines have the highest efficiencies, in the range of 30–40%, generally more than double an equivalent sized vertical axis system. They require relatively deep water for deployment and are often more expensive than other systems because the generators are submerged and there is complexity to their design. Horizontal axis turbines are usually self-starting with low cut-in speeds. Cut-in speed is the speed at which the turbine begins to turn and generate power. A tidal turbine will generate some power as it begins to spin and once it reaches the design speed it will produce the rated power. At this speed it will then offload thrust if the speed increases in order to manage the thrust loading. Turbines can do this by changing the pitch of their blades. This can be done with an active pitch system, a motor that turns the blades with respect to the flow, or passively using materials that will deform and change pitch with increased loading. The Schottel SIT turbine plot in Fig. 13 shows the 6.3 m rotor turbine cuts in at approximately 0.6 m/s, then once it reaches approximately 2.7 m/s, it begins to throw off thrust and it maintains the maximum power production until a maximum velocity of approximately 3.2 m/s. After this maximum speed, the turbine will cut out. The smaller diameter turbine will cut in slightly later and will continue producing power until a velocity of

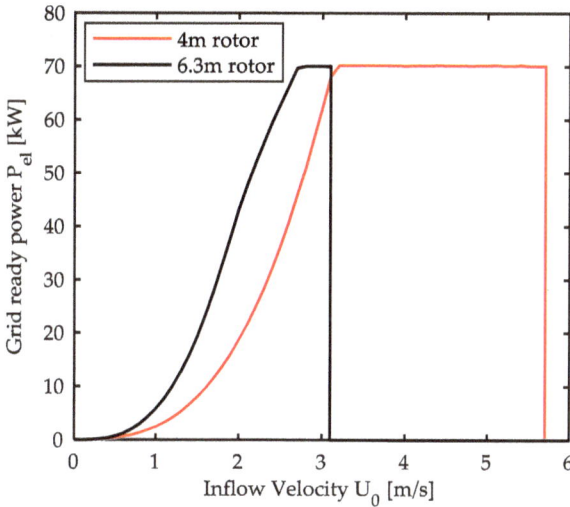

Fig. 13.   Power versus inflow velocity for the Schottel SIT Turbine, reprinted with permission.

approximately $5.7\,\text{m/s}$. The smaller diameter turbine is intended for faster flows. Slower flows require larger turbines (see Equation (6)). These turbines are unidirectional. In order for them to be able to produce power on both the flooding and ebbing tides, they need to rotate, pivot or have blades that can change the direction of the lifting surfaces.

A commercial vertical axis turbine that has been deployed globally is the New Energy Corporation turbine from Canada, as shown in Fig. 14.

Vertical axis turbines have very low efficiencies, 10–20%, but have versatility in application. They can be deployed in relatively shallow water and are significantly less expensive than the horizontal systems. This is due to lower complexity in blade design and because generally the generator is not submerged. Vertical axis systems are rarely self-starting and usually require external power to get them turning, but they do run at low cut-in speeds. Also, vertical axis turbines along with the cross-flow and screw styles being presented next are all able to produce power in flooding and ebbing tides.

The cross-flow style turbine looks similar to push style lawnmowers. Mavi Innovations, Fig. 15, have deployed this style in Canada and had some success.

The Archimedes screw style turbine has been adopted by a number of developers globally. There are completed and active projects that have tested this style in the Netherlands, the US and Canada. This style is often

Fig. 14. New Energy Corporation Vertical Axis River and Tidal Turbine, photo reprinted with permission (TBC).

Fig. 15. Mavi Innovations Cross-flow turbine, photo reprinted with permission (TBC).

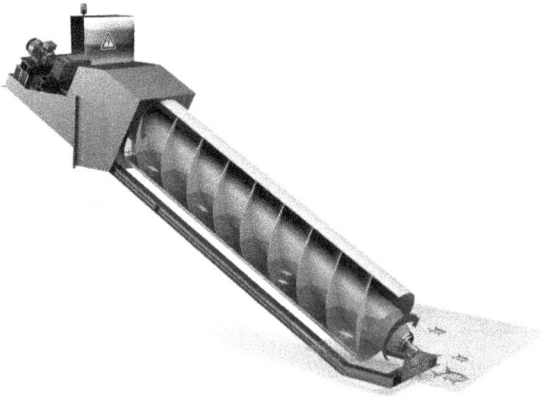

Fig. 16. The Fish Flow Innovations Archimedes Screw Turbine, photo reprinted with permission (TBC).

presented as "fish-friendly", where the expectation is that if a fish were to pass through the turbine, it would be unharmed. The design from Fish Flow Innovations shown in Fig. 16 is an example of this style.

12 Design Considerations

When turbine designers are developing their designs for specific applications, some fundamental considerations must be made.

While it may seem sensible to design for the maximum efficiency and maximum speeds to get the most value out of the turbine, it is important to design for the conditions that the turbine will see most often. When designing all turbines and engines, there is always a trade-off between efficiency and versatility of operation. If the design is focused on the high-speed conditions and the turbine sees those speeds for 5% of the time, then the turbine will not be optimized for 95% of the time that it is operating.

The current tables for a location can guide the designer for preliminary designs, see Fig. 17.

When developing a detailed design, higher resolution data at the site of interest must be acquired using acoustic current profilers, LiDAR radar or other means. Current data is plotted on a curve that can help to visually see the portion of the flow that is usable for power generation.

Designers then use the current velocity plot to build their designs. Using Fig. 18 as an example, it can be seen that a turbine with a cut-in speed in the range of 0.5 m/s would be needed at this site to produce power for half or more of the tidal cycle. The vertical lines cross the curves

GREAT BRAS D'OR AST Z+4   **2020**   CURRENT TABLES

January-janvier

Columns: **Turns / renverse** (Day/jour · Time/heure) — **Maximum / maximum** (Time/heure · Knots/noeuds)

Day / Jour	Turns heure	Max heure	Max noeuds
1		0022	+2.6
	0327	0648	-3.2
WE	1029	1326	+2.5
ME	1642	1916	-1.8
	2155		
2		0120	+2.2
	0409	0732	-2.9
TH	1109	1407	+2.4
JE	1733	2012	-1.7
	2257		
3		0222	+2.0
	0500	0824	-2.4
FR	1150	1447	+2.2
VE	1818	2108	-1.8
4	0016	0323	+1.8
	0604	0925	-2.0
SA	1230	1527	+2.1
SA	1858	2206	-2.2
5	0138	0427	+1.7
	0722	1025	-1.6
SU	1308	1609	+2.2
DI	1941	2303	-2.6
6	0253	0538	+1.7
	0840	1120	-1.4
MO	1346	1656	+2.5
LU	2029	2357	-3.1
7	0401	0651	+1.7
	0940	1208	-1.4
TU	1429	1745	+3.0
MA	2119		
8		0046	-3.4
	0459	0749	+1.8
WE	1025	1251	-1.6
ME	1518	1834	-3.4
	2208		
9		0128	-3.6
	0547	0832	+2.0
TH	1106	1334	-2.0
JE	1609	1922	+3.8
	2255		
10		0206	-3.7
	0624	0909	+2.4
FR	1150	1418	-3.1
VE	1701	2007	+4.0
	2339		
11		0243	-3.8
	0659	0943	+2.7
SA	1236	1502	-2.6
SA	1751	2053	+4.0
16		0030	+3.3
	0346	0641	-3.6
TH	1031	1318	+3.1
JE	1637	1914	-2.9
	2239		
17		0142	+2.9
	0443	0739	-3.2
FR	1122	1411	+3.0
VE	1734	2019	-2.9
	2359		
18		0251	+2.5
	0543	0842	-2.8
SA	1215	1509	+3.0
SA	1834	2128	-3.0
19	0123	0358	+2.2
	0651	0950	-2.5
SU	1309	1611	+2.9
DI	1932	2240	-3.2
20	0240	0515	+2.0
	0806	1057	-2.3
MO	1404	1716	+2.9
LU	2027	2349	-3.5
21	0352	0641	+2.0
	0918	1158	-2.2
TU	1458	1813	+3.0
MA	2118		
22		0048	-3.7
	0456	0747	+2.2
WE	1019	1252	-2.2
ME	1547	1902	+3.2
	2208		
23		0139	-3.9
	0550	0836	+2.4
TH	1109	1337	-2.3
JE	1633	1946	+3.4
	2256		
24		0227	-3.9
	0636	0915	+2.4
FR	1152	1418	-2.5
VE	1716	2028	+3.6
	2344		
25		0311	-3.8
	0713	0948	+2.4
SA	1229	1458	-2.7
SA	1758	2108	+3.7
26	0028	0348	-3.6
	0744	1016	+2.5
SU	1307	1538	-2.9
DI	1841	2147	+3.6

February-février

Day / Jour	Turns heure	Max heure	Max noeuds
1		0139	+2.0
	0421	0719	-2.3
SA	1038	1332	+2.5
SA	1703	2010	-2.3
	2340		
2		0240	+1.7
	0517	0808	-1.7
SU	1111	1410	+2.4
DI	1747	2106	-2.5
3	0105	0343	+1.4
	0623	0907	-1.2
MO	1144	1458	+2.5
LU	1842	2212	-2.7
4	0233	0501	+1.1
	0744	1014	-0.9
TU	1227	1555	+2.7
MA	1943	2321	-3.0
5	0353	0630	+1.1
	0859	1118	-1.0
WE	1326	1702	+3.1
ME	2044		
6		0019	-3.2
	0452	0730	+1.4
TH	0955	1217	-1.3
JE	1433	1808	+3.4
	2140		
7		0105	-3.5
	0530	0811	+1.9
FR	1045	1310	-1.9
VE	1542	1906	+3.7
	2231		
8		0144	-3.7
	0602	0844	+2.5
SA	1133	1359	-2.4
SA	1646	1957	+3.9
	2319		
9		0223	-3.9
	0634	0915	+3.0
SU	1219	1444	-2.9
DI	1744	2044	+4.0
10		0047	-4.1
	0709	0947	+3.4
MO	1303	1527	-3.4
LU	1838	2131	+4.1
11		0056	-4.2
	0744	1023	+3.6
TU	1346	1611	-3.4
MA	1930	2219	+4.0
16		0236	+2.3
	0522	0811	-2.4
SU	1127	1430	+2.9
DI	1756	2105	-3.3
17		0202	+1.5
	0633	0920	-2.0
MO	1222	1537	+2.7
LU	1858	2222	-3.3
18	0240	0511	+1.6
	0755	1031	-1.7
TU	1323	1652	+2.7
MA	1959	2343	-3.4
19	0353	0640	+1.7
	0909	1137	-1.7
WE	1425	1757	+2.9
ME	2059		
20		0051	-3.5
	0455	0740	+1.9
TH	1007	1231	-1.9
JE	1524	1849	+3.1
	2155		
21		0148	-3.6
	0543	0825	+2.1
FR	1053	1317	-2.2
VE	1616	1934	+3.3
	2246		
22		0235	-3.5
	0620	0858	+2.2
SA	1131	1358	-2.5
SA	1705	2015	+3.5
	2332		
23		0302	-3.4
	0648	0922	+2.7
SU	1206	1437	-2.8
DI	1751	2053	+3.6
24	0012	0324	-3.3
	0710	0941	+2.5
MO	1241	1516	-3.1
LU	1835	2130	+3.4
25	0047	0350	-3.3
	0729	1002	+2.8
TU	1316	1556	-3.2
MA	1917	2207	+3.2
26	0119	0419	-3.3
	0744	1027	+3.1
WE	1351	1635	-3.2
ME	1957	2243	+2.9

March-mars

Day / Jour	Turns heure	Max heure	Max noeuds
1		0103	+2.0
	0350	0626	-2.2
SU	0941	1231	+2.7
DI	1604	1912	-2.8
	2313		
2		0202	+1.5
	0436	0703	-1.6
MO	1006	1313	+2.7
LU	1652	2009	-2.8
3	0042	0306	+1.1
	0529	0752	-1.1
TU	1036	1406	+2.8
MA	1755	2120	-2.8
4	0218	0425	+0.7
	0647	0909	-0.8
WE	1127	1512	+2.8
ME	1904	2242	-2.9
5	0336	0602	+0.9
	0823	1041	-0.9
TH	1247	1635	+2.9
JE	2011	2349	-3.2
6	0423	0702	+1.4
	0931	1154	-1.4
FR	1416	1754	+3.2
VE	2113		
7		0037	-3.4
	0456	0740	+2.1
SA	1023	1251	-2.1
SA	1537	1855	+3.5
	2209		
8		0119	-3.6
	0529	0812	+2.7
SU	1110	1340	-2.8
DI	1645	1946	+3.8
	2301		
9		0202	-3.9
	0605	0842	+3.3
MO	1154	1423	-3.3
LU	1742	2033	+4.0
	2352		
10		0247	-4.1
	0643	0914	+3.5
TU	1235	1505	-3.7
MA	1833	2119	+4.1
11	0042	0334	-4.2
	0722	0949	+3.6
WE	1314	1548	-3.9
ME	1924	2208	+4.0
16		0225	+2.0
	0504	0744	-2.0
MO	1040	1357	+2.7
LU	1720	2045	-3.4
17	0113	0333	+1.5
	0617	0854	-1.6
TU	1139	1515	+2.5
MA	1828	2206	-3.2
18	0232	0500	+1.3
	0742	1006	-1.4
WE	1250	1633	+2.5
ME	1937	2335	-3.2
19	0342	0627	+1.5
	0855	1114	-1.5
TH	1405	1738	+2.7
20		0045	-3.3
	0437	0724	+1.8
FR	0949	1211	-1.8
VE	1512	1832	+2.9
	2141		
21		0139	-3.3
	0519	0804	+2.0
SA	1032	1257	-2.2
SA	1608	1917	+3.1
	2232		
22		0216	-3.2
	0549	0832	+2.2
SU	1107	1338	-2.6
DI	1659	1959	+3.3
	2315		
23		0234	-3.1
	0611	0848	+2.4
MO	1138	1417	-3.0
LU	1745	2038	+3.3
	2353		
24		0255	-3.1
	0629	0901	+2.7
TU	1209	1454	-3.4
MA	1828	2116	+3.2
25		0321	-3.1
	0648	0919	+3.0
WE	1239	1530	-3.6
ME	1907	2152	+3.1
26	0059	0350	-3.1
	0710	0941	+3.1
TH	1310	1604	-3.6
JE	1944	2227	+2.9

Fig. 17.   Current Table example for Great Bras d'Or.[13]

at 0.5 m/s and show the expected time that the turbine could produce power. It is also important to notice that the speed at which this particular turbine could operate at would be 0.5 m/s to ~1 m/s in the first flood and ebb tides and then would be 0.5 m/s to almost 2 m/s for the next two flood and ebb tides. This means that the chosen turbine should produce power from 0.5 m/s to 2 m/s to maximize the opportunity for generation at the site. Once these parameters are understood, then the turbine design can be chosen. A horizontal axis turbine with a large diameter and low cut-in speed could be used, or a vertical system may prove more versatile, depending on the other characteristics of the site including depth and access.

Other design considerations include foundation design, mooring systems, braking systems to slow down/stop the rotor, and maintenance of the system.

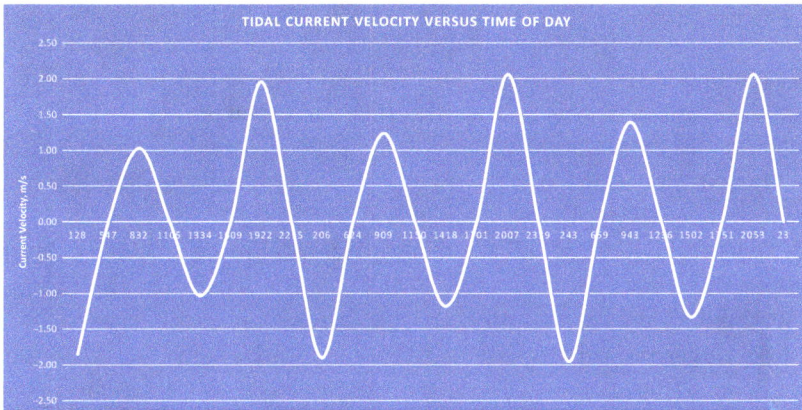

Fig. 18. Example of Current Velocity versus Time.

Where the system will be located will also highly influence these choices. Questions to ask regarding location include:

- Is the location deep enough for a large turbine and clearance? A horizontal axis turbine must have half the diameter distance at minimum of water above the top position of the blades to protect against sucking in air.
- Is the location suitable for an ocean floor mount? Is the seabed a material that can be drilled? Some locations have many meters of mud before any bedrock is encountered, which will eliminate drilling as an option. Is it financially feasible to drill in that location?
- Is there enough space for a turbine to swing on a mooring if it is required to swing to change direction as the tide changes? Some turbines are designed to pivot on their mounts, others are fixed and must rotate with a swinging mooring platform, and some have blades that switch direction when the tide changes.

13   Tidal Power Projects

Currently, tidal power is in the early stages of development. This means that many governments are creating incentives to encourage development of their sites. In Canada, developmental feed-in tariff programs have been successful in attracting projects. This means that the government approves a higher rate to purchase the power once it is flowing. This is intended to ensure that the extra developmental costs can be recouped over time, and reduces the upfront capitalization costs that can be very risky for governments to take on. Other countries use tax credit systems and combinations of grants and premium feed-in tariffs. In the UK, Renewable Obligation Certificates have been issued to ensure that the renewable power will be purchased at a premium rate. Countries have found different ways to encourage the development of marine energy, and these programs have stimulated the growth to date in the sector.

The development of full-scale in-stream tidal projects was led by the European Marine Energy Center, EMEC.org.UK located in the Orkneys in Scotland. They have supported the demonstration of more than 15 tidal developer projects to date and there have been many successes using their demonstration site. The Fundy Ocean Research Centre for Energy (FORCE) in the Bay of Fundy in Canada is the second site where multiple in-stream tidal power projects have been supported and demonstrated. There are a number of other sites around the world where technology is being further developed or one-off projects are being installed. The pace of development has been slow because the challenges are real. Yet interest has not waned, because of the value of a predictable renewable energy resource.

14   Socio-Economics

The introduction of a tidal project into a marine site and coastal community raises many questions among community stakeholders, as the technology is new and unfamiliar, and impacts are not fully understood.

Some of the concerns from community stakeholders include:

- line of sight;
- impact on other users of the site;
- tourism impacts;
- recreation impacts;
- environmental impacts;
- noise and activity;
- indigenous rights.

Positive aspects for a community include potential employment, community pride for supporting a new renewable energy technology and energy independence.

Tidal power technology is being developed at a time when engineers and developers are more cognizant of the potential environmental impacts of large projects. After many years of studying other forms of energy generation and their deployments, there is no reason to assume that a new renewable energy is going to be completely benign.

Many regions take an *adaptive management* approach that involves consultation of community stakeholders before projects are planned, updating stakeholders during the planning process, supporting the acquisition of baseline data on animal populations and environmental markers, and then monitoring the impact of the systems as they are deployed so that modifications can be made as needed.

In some regions, indigenous populations have treaty rights to the land and the path forward needs to include consultation by the federal government and inclusion of those communities in all stakeholder engagements.

Communities in the Orkneys and near the Bay of Fundy have been very supportive of the projects that were completed or are in development, although there have been concerns about impacts on fishing and line of sight. To date, there is no evidence that any projects on these sites have impacted negatively on the surrounding marine environment.

15 Other Tidal Power Systems

Other forms of tidal power system have been proposed but to date have had limited to moderate success in being demonstrated.

Kite systems have a kite with a turbine onboard that moves through the flow using the lift of the kite to move and a tether to contain it in space. (See Fig. 19). The tether is often also the cable that delivers the power from the turbine in the kite to the grid. Minesto from Sweden have had some success with this concept and have completed demonstrations.

The Venturi style system generates power by accelerating water through a choke point to create a pressure gradient and then the flow is used to run an in-built or on-shore turbine. The turbine can also be in the choke point, in which case a funnel or duct moves the flow through a smaller diameter pipe with a turbine blocking the flow. To date, it is not known if any systems have been brought to a commercial demonstration based on this technology.

Other system designs do not use a turbine. This includes *Flutter Vane Systems*, which are based on the principle of power generation from hydro-elastic resonance (flutter) in free-flowing water, and *Vortex Induced Vibration* (VIV) systems, which use vibrations resulting from vortices forming and shedding on the downstream side of a bluff body in a current. *Oscillating hydrofoils* use vertical oscillation of hydrofoils to produce

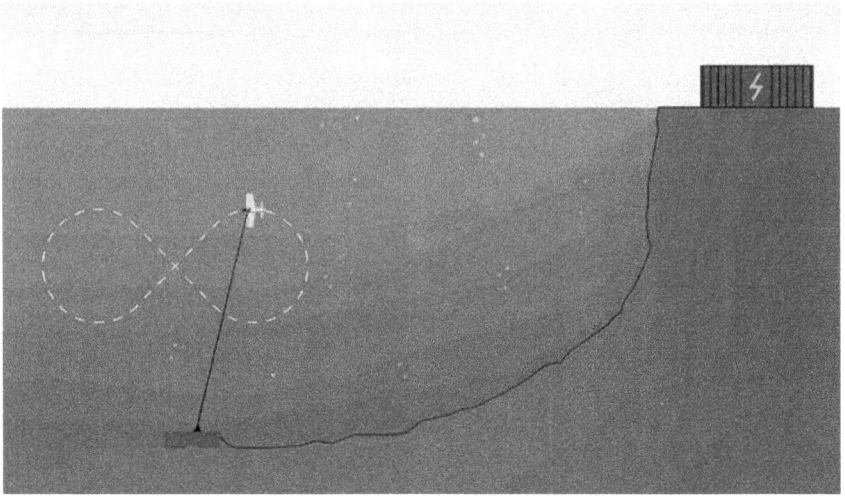

Fig. 19.   Minesto Kite Turbine[14] graphic reprinted with permission (TBC).

pressurized fluids, which are then used to operate a turbine. Finally, *Sails* that move linearly back and forth across a stream of water connect to a line and move a generator to produce power using the tidal flow.

A number of projects exploring these techniques for producing power have been conducted, but none have reached commercial success.

Another design similar to the sail is the *Kinetic Keel*. Essentially, it consists of a barge with a large keel that moves back and forth across a tidal flow, and the weight of the system pulls on cables connected to a generator. This system was tested at full scale and then scheduled to be put into operation by Big Moon Power but is now being reconsidered.

16   Commercial Turbines

Some turbines have been demonstrated and are near or at the commercial stage. Companies such as Andritz Hydro Hammerfest from Germany, Nova Innovation from the UK, Schottel from Germany, Atlantis Resources from Singapore and Verdant from the US have deployed systems in tidal sites and are working on the next stage of their projects. While setbacks have been common with many of these projects, in the past 2 years there has been an acceleration of demonstrable progress that bodes very well for the industry. For example, the Meygen project in the Pentland Firth, which uses Atlantis turbines regularly, tweets their power output—demonstrating the forward progress being achieved in tidal power.

17   Arrays

The next challenge of the industry, which has been demonstrated by Verdant and Atlantis and is being carefully considered in new projects going forward, is arrays. There are a number of direct challenges that are not yet well-understood and it is not clear how easily solutions to similar challenges on land will transfer to water. The interaction of wake fields from turbines and downstream turbines is being investigated; research to date has shown that the turbines need to be farther apart than initially anticipated to perform optimally (Fig. 20).

The staggering of turbines in the three-dimensional space of the ocean is currently being optimized using numerical models and smaller scale experiments. Other challenges for arrays include developing a system for interconnecting multiple turbines electrically in a site, operating and

Fig. 20.   Graphic depicting Meygen Array,[15] graphic reprinted with permission (TBC).

maintaining multiple turbines in a site and anchoring vessels required for repair work in a site with multiple turbines.

18   River Turbines

River turbines are very similar to tidal turbines; in fact, tidal turbines can often be used in river flow. Still, there are some differences that need to be considered. For example, river turbines often do not need the durability required for ocean conditions. Indeed, some attempts to operate river turbines in the ocean have resulted in redesigns to address durability. Table 2 highlights typical differences between river and tidal turbines.

19   Ducts

Ducts are add-ons to turbines that can be used to direct flow, change flow, increase the speed and more. Some ducting is used simply to prevent debris from entering the turbine through blocking or diverting. Examples of the kinds of ducting that can be added to vertical and horizontal axis turbines are included in Fig. 21. To date, ducting is not commonly found on horizontal axis turbines or even vertical axis turbines. It is possible that the delays caused by debris being diverted into the duct — requiring a visit to the turbine to clean it — outweigh the benefits. Ducting has been used in the Archimedes screw design from the Netherlands; it is used to provide a channel for fish to travel in and to accelerate the flow over the screw.

20   Deployment and Maintenance

For any tidal project, marine operations are one of the most significant challenges. This means that plans for both deploying turbines and maintaining them need to be considered right from the beginning of the design of the project.

Table 2.   River and Tidal turbine characteristics.

River	Tidal
Power usually 5–100 s of kW	Power 500 kw to MWs
Flow unidirectional	Flood and ebb
Seasonal and daily flow changes	Diurnal or semi-diurnal flow changes
Water density is fairly stable	Density and salinity change with temperature
Dynamic control systems	Passive controls
Usually stand alone	Grid connected

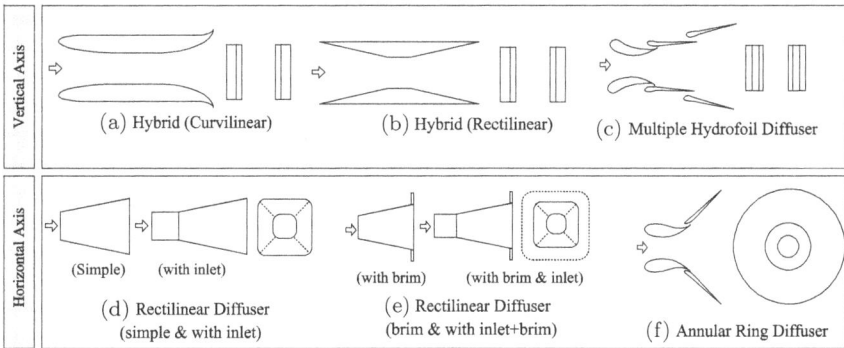

Fig. 21.   Duct styles.[16]

The primary challenge is operating in the flow. The same characteristics of a tidal site that make it attractive make it very challenging to build in. Vessels that deploy turbines need to be able to hold station during lowering and lifting operations. All these vessels have limits to the speeds they can operate in, the higher the speed, the more expensive the vessel and the fuel required to maintain the station.

Looking at the current plot in Fig. 18, it can be seen that there are approximately 1–2 hours every time the tide changes direction when the velocity is less than 0.5 m/s. In some sites, such as the Minas Passage in the Bay of Fundy, there can be less than 2 hours when the flow is under 2 m/s and where flow never really goes below 1 m/s. The vessels required to operate in the faster sites will dictate the project team's ability to affordably complete their operations (Fig. 22).

This challenge to operate in high flow means that often operations are planned for the neap tide and happen in the short window of lower flow. This can mean that 6 hours of operations can take 2–3 days, depending on whether the teams work at night. Safety is an important consideration.

These time restrictions also apply to drill ships that are needed to install piles for floor-mounted turbines or for mooring systems. Innovations in subsea drilling are being developed specifically for this industry. For example, one concept being considered is to have a tetrahedral frame that can be left on the ocean floor with an umbilical cable that can be connected to a drill ship when the ship is able to maintain station on site. This could reduce the cost of drilling substantially and make drilling a viable option in some very high flow sites. This is the kind of innovation that is needed to make tidal power possible. The next step is to make it financially feasible.

Fig. 22.   Deployment of the Schottel turbine and Sustainable Marine Energy platform project in the Bay of Fundy, photo reprinted with permission.

The flow speed will then of course also affect maintenance. The ability to lift a turbine, move a turbine or operate in a field of turbines is directly impacted by the high flow in the field.

There is an expectation in the industry that once arrays are deployed, operation specific vessels can be built directly for the industry and bring down costs significantly.

This industry has been building a wealth of knowledge about deployment, operation and maintenance in tidal sites over the past 15 years. Some of this learning has held up well while other knowledge has been improved upon and updated over time. For example, one topic of focus in the late 2000s was the idea that large trees and blocks of ice would be moving through tidal sites after being eroded away by the high flow and storms. There were many discussions about whether turbines would be able to survive the impact. To date, this has not been a concern for deployed projects, and sites are being monitored for material moving through sites. It is now not expected that this will be a major concern for the safety of turbines. Turbines are low enough in the water column to not be hit, and ice will either float or sink if it is embedded with rocks/debris. There is not much concern about neutrally buoyant ice, as there is little evidence to suggest that it is present in significant amounts.

21 Modeling and Testing

Given the challenges of operating in real-world deployment sites, testing systems in the ocean is a very expensive proposition. The world of ocean engineering has a long history of applying modeling and experimentation and this expertise is being regularly applied to tidal projects.

Projects are often run through Computational Fluid Dynamics (CFD) simulations to understand the performance and loading on the systems. Controlled experiments are also used extensively for both understanding and optimizing the systems.

Towing tanks have been used for over 100 years in Naval Architecture to study the performance of ships, and they are used now to study turbines. Most tow tanks are similar to that shown in Fig. 23; this one is 54 m long,

Fig. 23. Memorial University of Newfoundland Towing Tank, photo reprinted with permission.

Fig. 24.   FloWave facility at the University of Edinburgh.[17]

while the largest tanks are 200 m or more long. The tanks have carriages that can tow the turbine at speeds as high as 10 m/s in the large tanks. Turbines are tested at small scales and optimized before being built at full scale. The larger the tank, the larger the model turbine that can be tested. Ideally, because water doesn't scale, turbines should be scaled at 10 times or less to get the best quality data.

Platforms with turbines are tested in ocean testing basins that can simulate the open ocean conditions so that expected and extreme loads and forces on the platform and mooring system can be measured. Fig. 24 shows the FloWave Ocean Energy Research basin at the University of Edinburgh. This is a state-of-the-art facility that can produce flow in every direction.

22   The Future of Tidal Power

Tidal power has been slow to meet its potential at full scale because of the capital costs, the cost of deployment and, as a result of limited projects, the uncertainty around maintenance and operation costs. It has been said over and over again in this community that projects need to tackle low-flow sites before going to the very high-flow sites. The counter argument has always been that high flow is simply an engineering problem and it is possible to go straight to the high-flow sites. While high flow can be tackled with quality engineering, the cost of deploying needs to be covered and the financial backers of these projects need to have some assurance of what tidal power technology can offer financially. As a sector, tidal power needs more energy producing sites connected to the grid in lower flow regimes and then with time we can prove the reliability of this technology to the financial backers and insurers of tidal power. There is a real opportunity to develop this predictable and reliable renewable energy source. Grid connected projects will help immensely in building the trust needed to deploy in high-flow sites. Tidal power is a valuable resource for many coastal, end of the power line communities, offering energy security and employment as well as contributing to reduced greenhouse gasses.

References

1. https://oceanservice.noaa.gov/education/tutorial_tides/tides02_cause.html.
2. https://www.britannica.com/science/Earth-tide.
3. https://www.waterlevels.gc.ca.
4. https://tidesandcurrents.noaa.gov/images/restfig6.gif.
5. https://oceanservice.noaa.gov/education/tutorial_tides/tides07_cycles.html.
6. https://manoa.hawaii.edu/exploringourfluidearth/media_colorbox/2310/media_original/en.
7. https://www.collectionscanada.gc.ca/eppp-archive/100/200/301/ic/can_digital_collections/west_nova/tidal.html.
8. https://www.novascotia.com/see-do/attractions/annapolis-tidal-station/1329.
9. B.K. Hodge (2010). Alternative Energy systems and applications, Wiley, US.
10. https://svs.gsfc.nasa.gov/stories/topex/index.html.
11. R. Charlier and J. Justus (1993). *Ocean Energies: Environmental, Economic and Technological Aspects of Alternative Power Sources, Ed. 1.* Amsterdam: Elsevier.
12. https://science.howstuffworks.com/environmental/earth/oceanography/deep-ocean-exploration.htm.
13. https://tides.gc.ca/eng/data/table/2020/curr_ref/620.
14. https://minesto.com/our-technology.
15. http://www.scottishenergynews.com/meygen-developer-of-the-worlds-largest-tidal-energy-project-opens-new-uk-head-office-in-scotland/.
16. M.J. Kahn, G. Bhyuan, M.T. Iqbal, J.E. Quaicoe (2009). Hydrokinetic Energy Conversion Systems and assessment of horizontal and vertical axis turbines for river and tidal applications: A technology status review. *Applied Energy*, Elsevier, V86, pp. 1823–1835.
17. https://www.flowavett.co.uk/home.

General References

   i. C. Bates (2008). Tidal power feeds electricity to National Grid in world first, 17th July http://www.dailymail.co.uk/sciencetech/article-1035978/Tidal-po wer-feeds-electricity-National-Grid-world-first.html.

  ii. R. Karsten, H., McMillan, J.M., Lickley, M.J., and R.D. Haynes (2008). Assessment of tidal current energy in the Minas Passage, Bay of Fundy. *Journal of Power and Energy, IMechE*, Vol. 222, Part A.

 iii. CEA, Power Generation in Canada, 2006, Canadian Electricity Association, www.electricity.ca.

 iv. Atlantis Strom http://www.atlantisstrom.de/description.html.

  v. C. Wessner and C. Bear, (2009). Vertical Axis Hydrokinetic Turbines: Practical and Operating Experience at Pointe du Bois, Manitoba, New Energy Corporation Inc., Calgary, Alberta, Canada.

 vi. http://www.newenergycorp.ca/LinkClick.aspx?fileticket=7GLwd1%2BEqh 4%3D&tabid=84&mid=471.

 vii. EMEC, European Marine Energy Centre, http://www.emec.org.uk.

About the Editor

Gerard "Gary" Crawley is the President of Marcus Enterprises LLC based in North Carolina. Previously, Professor Crawley served as the Director of the Frontiers Engineering and Science Directorate of Science Foundation Ireland from 2004–2007. Prior to this, Professor Crawley was the Dean of the College of Science and Mathematics at the University of South Carolina from 1998 to 2004. At Michigan State University, he was Dean of the Graduate School from 1994–1998 and earlier Chair of the Department of Physics and Astronomy from 1988–1994. Professor Crawley served two terms at the US National Science Foundation, one as the Director of the Physics Division, 1987–1988, and earlier as a Program Officer in the Nuclear Physics Program. He has also served as the Chair of the Nuclear Physics Division of the American Physical Society in 1991–1992. Dr. Crawley was born in Scotland, but his first degree is from the University of Melbourne in Australia. He obtained his PhD in Physics from Princeton University, USA, in 1965. He is the author of over 150 articles in refereed journals and a text book titled *Energy* published in 1975. He is also the Editor of the *World Scientific Handbook of Energy* published in February 2013. Currently, Professor Crawley is the editor of the World Scientific Series on *Current Energy Issues*. The first five volumes in the series have been published already. He is also the co-author of a book, *The Grant Writers Handbook,* which was published by Imperial College Press in 2015. He has previously consulted for the National Research Foundation of the UAE (2008–2011) and the National Center for Science and Technical Evaluation, Republic of Kazakhstan, from 2008–2012.

About the Contributors

Giovanna Cavazzini (Chapter 5) received her degree with honors in Mechanical Engineering in 2003 from the University of Padova and her doctorate with the European Berlin Energetics program in 2007 from the University of Padova. In 2005, she spent a research period at the Ecole Nationale Superieure Arts et Métiers Paris Tech de Lille with emphasis on the study of unsteady turbulent phenomena developing in turbomachines. In 2009, she received the qualification of Maitre de Conference from the Ministère de l'Enseignement Supérieur et de la Recherche in France and since 2016 she has been an Associate Professor of Fluid Machines and Energy Systems at the University of Padova. Her main research interests include analysis and modeling of renewable energy systems, techno-economical optimization of hydropower plants and design optimization of fluid machines with particular emphasis on hydraulic turbines and pump-turbines. She is the author of more than 90 scientific publications, the most part of which published in International Journals with Impact Factors and in Proceedings of International Conferences. She is Deputy Chair of the joint sub-program of the European Energy Research Alliance (EERA) on "Mechanical Energy Storage".

Magnus Gehringer (Chapter 4) is CEO of Consent Energy LLC with offices in Washington, DC, USA, and Reykjavik, Iceland. He is an internationally recognized expert in the use of geothermal resources for power generation and direct/residual heat uses. Magnus covers technical, financial and policy issues, including project development and business development, business strategy, project feasibility and project finance. He has work experience in over 40 countries. Over a period of six years, Magnus was Senior Geothermal Energy Specialist at the World Bank/ESMAP and consultant to the IFC, where he worked on development of geothermal projects in many countries in Africa and several in Central America and Asia. Prior to joining the World Bank, Magnus served as Executive Director of Business Development at the National Power Company of Iceland (Landsvirkjun). Until 2007, he was Chief Executive Officer of Exorka Ltd., an Icelandic engineering company that specialized in the design, construction and operation of "Kalina" binary power plants for geothermal, biomass and industrial waste heat recovery application. Magnus Gehringer is also a public speaker and author on issues related to renewable energy, technological solutions, energy project finance, politics and risk analysis of energy projects. He has authored multiple books and articles including the "*Geothermal Handbook*" published by World Bank in 2012, which is still a leading guide to the global geothermal industry, and the chapter on geothermal energy in the "*World Scientific Handbook of Energy*" published in 2013. Magnus holds an M.Sc. degree in Business Economics and a technical diploma degree, both from Iceland.

Sue Molloy, PhD, P.Eng. (Chapter 8) is President of Glas Ocean Electric and Adjunct Professor of Ocean Engineering at Dalhousie University. Dr. Molloy specializes in Electric Boats & Ships, Marine Renewable Energy (MRE) and Sustainable Ocean Engineering. Dr. Molloy served as an elected member of the Council for Engineers, Nova Scotia. She is the former Co-chair of the Canada–China Track II Energy Dialogue MRE subcommittee, is the international chair for the International Electrotechnical Committee (IEC) River Turbines Project team and is a former

Board Member of Marine Renewables, Canada. Dr. Molloy recently received the IEC 1906 international award for major impact in furthering standardization and related activities in electrotechnology. She recently joined ISO TC 8 Ships and Marine Technology as a Canadian Expert and is a Member of the Canadian Forum on Marine Autonomous Surface Ships (MASS). Dr. Molloy is a former Board Member of FORCE and a past President and General Manager of Black Rock Tidal Power, now Sustainable Marine Energy Canada.

Torbjørn K. Nielsen (Chapter 5) obtained his Master's degree in 1976 and his PhD in 1991 from the Norwegian Institute of Technology, Trondheim, Norway. He has been a Professor at the Waterpower Laboratory, the National Technical Norwegian University (NTNU), since 2002. During his time at NTNU he has supervised 11 PhD students. He retired from this position in 2020. His previous experience was as a Senior Researcher at SINTEF until 1995. In addition, he was involved with turbine testing and design where he was employed by Kværner Energy and later GE Energy. He has worked his whole professional life with hydraulic machinery and systems.

Gérard C. Nihous (Chapter 6) graduated from the École Centrale in Paris in 1979 and from the University of California at Berkeley in 1983. His doctoral thesis in Ocean Engineering dealt with wave power extraction. After moving to Hawaii in 1987, he became involved in research on Ocean Thermal Energy Conversion (OTEC) for more than a decade. He taught a graduate course on renewable energy at Hiroshima University in 1996 and 1997. Gérard also worked extensively on the ocean sequestration of CO_2 until 2002. He was hired by the University of Hawaii's Hawaii Natural Energy Institute in 2003, primarily to do research on methane hydrates. He later joined the Department of Ocean and Resources Engineering in 2009, where the focus of his activities was, once again, the promotion and development of marine renewable energy. He retired in 2019.

Pål-Tore Storli (Chapters 5) is an Associate Professor at the Waterpower laboratory at the Norwegian University of Science and Technology (NTNU). Born in September 1980, P.-T. Storli obtained his M.Sc in Energy and Environmental Engineering in 2006, and his PhD in Fluids Engineering in 2010, both at NTNU. He has been working extensively at the interphase between natural sciences and technology, especially regarding the technical aspects of hydropower and pumped storage and the environmental impacts of such technologies and their operation. P.-T. Storli has been involved in the structuring and alignment of European research efforts on hydropower involved in the establishment of a Joint Programme on Hydropower within the European Energy Research Alliance (EERA), where he has been Vice Coordinator.

Arne Vögler (Chapter 7) initially trained as an Electronics Engineer in Germany, before diversifying towards Mechanical and Energy Engineering in the UK. He obtained a PhD on the *Wave characterisation for high and medium energy sites in Scotland* with the University of Aberdeen, and previously graduated with a research Master's degree on *the use of hydrogen in marine applications*. As a Chartered Engineer and Chartered Marine Engineer, Dr. Vogler has spent the last ten years leading a research group at the University of the Highlands and Islands, investigating and applying innovative wave and current data acquisition approaches in high energy environments. He has authored and co-authored a wide range of journal articles and conference proceedings, and has been a frequent speaker at numerous engineering conferences in the Americas and Europe. He has been Principal Investigator and Co-Investigator of a number of national and international research projects, and has also organized a high-impact international conference on the environmental interactions of marine renewable technologies. As commercial and operations director of Hebrides Marine Services Ltd, a company founded by him, Dr. Vogler successfully delivered a range of projects for commercial and public sector clients in the marine energy sector, often in challenging environments and circumstances. Currently, Dr. Vogler is the lead engineer of the Marine Energy Engineering

Centre of Excellence, where he is supporting the development of engineering solutions for marine energy in Wales as part of the UK Offshore Renewable Energy Catapult (OREC).

Li Zhang (Chapters 1, 2 & 3) received her PhD degree in 1985 from Oxford University and was a Research Fellow at Oxford University, UK. She was then a Lecturer at the University of Bradford and is currently an Associate Professor in the School of Electronic and Electrical Engineering at the University of Leeds, UK. Dr Li Zhang has been an Adjunct Professor in Chongqing University, China, since 2006. She is also a Joint Grant holder in the China State Natural Science Foundation Fund (60712, 2014 01-2017 12). She is an Associate Editor for IEEE, Transaction on Power Electronics and has also been the Associate Editor of IET proceedings on Power Electronics between 2014 to 2017. She has authored and co-authored three books on power converter circuits and wind power electricity generation. She has authored and co-authored more than 100 technical papers in the fields of power electronics, renewable power generation system and wind generator control.

Qiou Yang Zhou (Chapter 3) did his MSc in Engineering Technology and Business Management, at the University of Leeds during 2017/18. He is currently doing a PhD degree in the area of Vehicle Operation Engineering, at Southwest Jiaotong University, since 2019. His research topic is on reliability analysis and fault diagnosis for high speed trains.

Index

A

a lifting force, 13
AC-DC-AC converter, 53–54
acoustic current profilers, 251
active stall, 75
active stall control, 21
adaptive management, 255
aerodynamic controls, 19
aerodynamically formed rotor blades, 2
air density, 8
air-gap power, 59
airstrip turbulence, 40
Albatern, 214
Albatern's WaveNET, 220
angle of attack, 14–15, 20–21
angular momentum, 154–155
angular momentum equation, 154
animal, 255
annual averaged wave power, 207
Annual Energy Production (AEP), 242
aquaculture, 221
AquaHarmonics, 219
Aquamarine, 201, 230
Aquamarine Power, 211, 222
Archimedes Waveswing, 219, 247
armature windings, 42
aromatic polyamide, 80
arrays, 257
artificial upwelling, 187

asynchronous machine, 40
Australian Wave Energy Research Centre (WERC), 220
average load factor, 75
average power rating, 67
average total installed costs of onshore wind farms, 70
AW-Energy, 213
AWS Ocean Energy, 219
axial forces, 15

B

back pressure, 98
ball screw generator, 216
barrages, 240
basalt fibers, 80
base-load, 97–98, 173, 184
bathymetric, 239
bathymetry, 235
Bay of Fundy, 239
Bernoulli equation, 146–147
Bernoulli's principle, 13
Betz Law, 11, 13
Betz Limit, 245
binary, 98, 101
binary set, 142–144, 146
binary units, 143, 145
biological impact, 227
blade pitch angle, 15, 31
blocking, 258
Bloomberg New Energy Finance, 67

273

www.ingramcontent.com/pod-product-compliance
Lightning Source LLC
Chambersburg PA
CBHW050545190326
41458CB00007B/1922